I0119940

Dingo Bold

ANIMAL PUBLICS

Melissa Boyde, Fiona Probyn-Rapsey & Yvette Watt, Series Editors

The Animal Publics series publishes new interdisciplinary research in animal studies. Taking inspiration from the varied and changing ways that humans and non-human animals interact, it investigates how animal life becomes public: attended to, listened to, made visible, included, and transformed.

Dingo Bold

The Life and Death of K'gari Dingoes

Rowena Lennox

SYDNEY UNIVERSITY PRESS

First published by Sydney University Press
© Rowena Lennox 2021
© Sydney University Press 2021

Reproduction and communication for other purposes

Except as permitted under the Act, no part of this edition may be reproduced, stored in a retrieval system, or communicated in any form or by any means without prior written permission. All requests for reproduction or communication should be made to Sydney University Press at the address below:

Sydney University Press
Fisher Library F03
University of Sydney NSW 2006
Australia
sup.info@sydney.edu.au
sydneyuniversitypress.com.au

A catalogue record for this book is available from the National Library of Australia.

NATIONAL
LIBRARY
OF AUSTRALIA

ISBN 9781743327319 paperback
ISBN 9781743327326 epub
ISBN 9781743327371 mobi
ISBN 9781743327388 pdf

Cover design by Mary Stanley and Miguel Yamin
Cover image by Neil Cambourn © Queensland Government, reproduced with the permission of the Department of Environment and Science, Queensland
Map by Laurie Whiddon
Internal photographs in Prelude © Rowena Lennox; in Chapter 4 © Jennifer Parkhurst; and in Chapter 7 © Dan Novak

For Mum and Zefa

Contents

Map of K'gari

Scale 1:678,000

0   10   20
Kilometres

Sandy Cape
Sandy Cape Lighthouse

H e r v e y

B a y

Marloo Lakes
Rooney Point
Ngkala Rocks
North Ngkala Rocks
South Ngkala Rocks

Orchid Beach
Waddy Point
Connors Corner
Champagne Pools
Indian Head

Wathumba

GREAT
SANDY
NATIONAL
PARK

Awinya

Woralie

Wyuna Creek

Dundubara

Gungul Point

Lake Allom

Cathedral Beach
K'gari
The Pinnacles
SS Maheno wreck

Moon Point
Point Vernon
Sandy Point
Eli Waters
Blacktown Point
Woody Island Lighthouse
Little Woody Island

S o u t h

P a c i f i c

O c e a n

Hervey Bay

Woody Island

Kingfisher Bay Resort Village

K'gari
(Fraser Island)

Happy Valley
Yidney Rocks
Winnam

River Heads

Poyungan Rocks
Gabala

Wangoolba Creek barge landing

Lake McKenzie
Lake Wabby
Central Station
Cornwells
Lake Wabby Beach
One Tree Rocks
One Tree Rocks

Second Valley
Eurong
Wongai

Lake Boomanjin

Govi
Geranweena Creek
Dilli Village

Maaroom

G r e a t

S a n d y

S t r a i t

Boonooroo
Tuan
Poona

GREAT
SANDY
NATIONAL
PARK

Snout Point

Tinnanbar

Jabiru Swamp
The Spit

Coplantoi Creek
Wide Bay Harbour

Hook Point
Inskip Point

Burnett Heads
Bargara
Bundaberg
Elliott Heads

H e r v e y

B a y

Howard
Torbanlea

Hervey Bay

River Heads

Maryborough

Urangan
K'gari
(Fraser Island)
Happy Valley
Kingfisher Bay Village Resort

Sandy Cape

Waddy Point
Indian Head

GREAT
SANDY
NATIONAL
PARK

Eurong

Dilli Village

Tiaro

Tuan
Poona

S o u t h

P a c i f i c

O c e a n

Tin Can Bay

Rainbow Beach
Double Island Point

Hook Point
Inskip Point

Gympie

ix

Prelude

... drums rolled off in my forehead
And the guns went off in my chest
    The Triffids, 'Wide Open Road', 1986

He has a multicoloured tag in his ear.

Blood throbs at the back of my head as he comes closer. My heart is beating so hard and loud I am sure he can hear it.

He looks up at me. He seems to be asking for something. The way his torso narrows to his rump looks skinny from where I stand. He flinches, almost imperceptibly, and I think, *He's as nervous as I am. I can scare him off if I need to.*

But I stay still.

He goes around behind me. I look straight ahead. He walks away.

Later I find the photos I took on my phone. Initially I don't know why I took a picture with a foreground full of sand, tracked with light brown roots and wispy pale green strokes of dune grass; a middle ground of silhouetted casuarinas on the dune; and a cerulean sky washed by wave-like clouds. Only when I zoom in on the centre of the image do I notice, in the shade cast by the dune, a dingo heading straight for me. Another dingo, also hardly distinguishable in the shadow, is in the picture too, further away on the dune with his back toward me.

I took one more photograph as he approached.

I wish I had an image of his face when we conversed but photographing him would not have been conversing with him. I have no memory of taking two photos of him leaving, his shadow in the afternoon sun so much bigger than he is.

1
Wildlife

when a dingo is hungry it will kill to eat and then it takes it
   Barbara Tjikadu in Lindy Chamberlain, *Through My Eyes: An Autobiography*, 1990

There was never a time when dingoes didn't exert an illicit pull in my mind. In the 1970s when I was growing up in an outer suburb of Sydney, people were not allowed to keep them as pets. They had a reputation for being impervious to human control, a law unto themselves – qualities that appealed to me. Our family had blue cattle dogs, who counted dingoes among their ancestors. Our dog Beau could beat every other dog in the street in a fight and jump unbelievable heights. No enclosure could contain him. When my parents separated, he – not we – decided which household he would live in. His intelligence, prowess and independence might have come partly from his dingo genes.

Three of our dogs died like dingoes when five-year-old Possum and her two grown pups took a strychnine bait in the woodpile on the verge outside my father's house. A neighbour told us how Possum had watched her pups die before her.

I knew that, historically, dingoes were considered pests – I read *Dusty*, a novel written in the 1940s about an outlaw dingo–kelpie cross[1]

---

1   Davison 1983

when I was a child – but until a few years ago when I started researching
dingoes as part of an essay I was writing about my kelpie–cattle dog
Zefa, I was unprepared for how vehement emotions around dingoes
are. On the radio I listen to farmers from around Blackall in
Queensland rail against the wild dogs that are driving them out of
sheep production. I hear the rage and powerlessness people feel when
they find their wounded animals after attacks. I see pictures of dingoes
hanging from trees and strung up on fences by their back legs, their
muzzles stretched toward the earth. On the internet are reports
produced by state governments and organisations such as
WoolProducers Australia and the Invasive Animals Cooperative
Research Council that describe wild dogs and dingoes in
well-researched detail and indicate more effective ways to 'manage'
them. 'Manage' is a euphemism for eradicate. This violence is part of the
shadow story of my country.

Here's trouble, I think. *I'm in.*

In moments of lesser bravado I wonder why we go around killing
other living beings on this scale.

I start searching for dingoes. My husband, children and I take a few
days to drive from Sydney to the Western Plains Zoo in Dubbo. On the
way we go for bushwalks and I look for dingo scats and tracks in the
Blue Mountains and at Mount Canobolas in the central west of New
South Wales. I picture dingo habitat on the scrubby undulating edges of
the Great Dividing Range.

At the zoo in Dubbo I stand at the dingo enclosure for two days
and watch Bunjil and Mojin go about their business and doze in the
sun.[2] One of their keepers tells me Bunjil, the male, is named for
the eagle ancestor and creation spirit of the Kulin Aboriginal people
from central Victoria. Mojin, so the Ngulugwongga Aboriginal story
from Daly River in the Northern Territory goes, was a male dingo
who had collected some yams to eat. He couldn't make fire to cook
the yams so he tried to steal a firestick from some women camped
nearby. But they hunted him away. So then his friend Little Chicken
Hawk (Djungarabaja) went to the women's camp. He was small and the
women didn't see him. As he stole a glowing ember he cried out 'Diid

---

2   Lennox 2014

2

… Diid!' before he flew back to his own camp to start a fire. But Mojin couldn't wait. He had already eaten his yams. Now, unlike Chicken Hawk, Mojin cannot talk. And he eats his food raw.[3]

My family and I visit a wildlife park at Helensburgh on the southern outskirts of Sydney. I go to the dingo enclosure to watch the dingoes and talk to the keeper who comes to feed them. My children are impatient. I visit a dingo breeder in Victoria. The pups undo my shoelaces and climb up my legs. Their fur is soft and they smell of ammonia. At a conference I touch noses with a rescue dingo. She licks my mouth.

I start to interpret our dog Zefa's behaviour in the light of what I am learning about dingoes' ecological relationships. At night when Zefa chases a possum down the driveway and urinates at the base of the tree it has climbed I deduce that she is communicating with the possum: *I'm here. Come down and face the consequences.* When she eats all of the food in our friends' cat's bowl, perhaps she's not just being a glutton, perhaps she is performing an instinctive duty, showing the cat, a mesopredator, that she is the apex predator, top of the food chain: *What's yours is mine.* My daughter tells me that in every conversation I talk about dingoes. Sometimes my children tolerate my obsession; sometimes they are bored with it, frustrated by it.

We go to Uluru. The only dingoes we see are on signs at the campground warning campers to keep their shoes in their tents. There's what might be a canid-sized scat and an indistinct paw print near the boardwalk on a lookout between Kata Tjuta and Uluru. But, for me, it's all about dingoes. Kurpany – the giant shape-shifting *mamu* (ghost or evil spirit) dingo, who came from the west, attacked and killed the Mala men and left his footprints on the Rock as he chased the surviving Mala people hundreds of kilometres to the south and east – could be anywhere. Or everywhere.[4] Is that warm westerly rippling the rapiers of yellow spinifex after sunrise Kurpany? Is the crest of that dune – rising like a red sand wave and echoing the shape the Rock – the one he stands on, watching, salivating and grinning in anticipation before he careers across the plain to his kill?

---

3   Berndt and Berndt 1977, 395–6
4   Director of National Parks 2012, 111; Layton 1989, 5–7

We cycle around the Rock. Obliviously we pass a place of rocks and sandy patches at the south-west corner, between Mutijulu waterhole and the Mala car park (formerly the starting point of the climb), where, below a lichen stain on the rock face, tourists found the clothing of baby Azaria Chamberlain a week after she went missing from her tent in the campground four kilometres away on 17 August 1980. Aboriginal trackers found dingo and puppy tracks near a den about thirty metres west of where the clothes were found, and adult dingo tracks ten metres east of the clothes.

Unlike Kurpany, who left the splattered remains of the Mala men that he killed on the north face of the Rock, the dingoes who took Azaria were not so wasteful. Her body was never found.

Her white stretch towelling jumpsuit was sitting on its back,[5] concertinaed,[6] and soiled around the neck with blood. Whether Azaria's Bonds ribbed singlet was inside the jumpsuit or protruding from it was a matter of dispute because neither the family who found the clothes nor Senior Constable Frank Morris – the police officer in charge at the Rock, whom they alerted – photographed them before Constable Morris moved them. Nearby was Azaria's disposable nappy, torn, with the insides exposed. Her tiny white booties, knitted by her grandmother with special hand-twisted ties and ribbed up the leg with moss stitch at the top of the toe, were still inside the jumpsuit, fleshing out the spot where her feet would have been. The ground around had been disturbed and a patch of flattened undergrowth indicated an animal had lain down.[7]

Ayers Rock Chief National Parks Ranger Derek Roff and his colleague Ranger Ian Cawood had not known about the den with its recent litter near where the clothes were found. The nearest one they were aware of was one kilometre east.[8] Morris shot a lactating female in the vicinity a week later.[9] Other dingoes from the den were shot and the contents of their stomachs examined, revealing no sign of human

---

5   Chamberlain 1990, 81
6   Young 1989, 240
7   Morling 1987, 240
8   Morling 1987, 239
9   Young 1989, 247

bone or protein.[10] Off-the-record reports claimed rangers shot over sixty dingoes after Azaria was taken but the official tally was only 'half a dozen or so'.[11] In 1986 Barbara Tjikadu, an Indigenous custodian of Uluru, told the Morling Royal Commission of Inquiry into the Chamberlain Convictions that there 'used to be a lot of dingoes there'.[12]

There was much debate in the courtrooms about whether dingoes would leave clothes the way Azaria's had been found. Then as now, dingo experts did not agree. Ranger Derek Roff believed 'different dingoes might leave the clothes differently arranged'. He said, 'it could be scattered, and sure, it could be exactly like this'. He 'was not prepared to hazard a guess as to how a dingo might have left the clothes'.[13]

In 1986, a search for the body of a tourist who had fallen from the Rock uncovered Azaria's famous white knitted matinee jacket – which her mother, Lindy, insisted she had been wearing on the night of her disappearance – about 150 metres away from where her jumpsuit was found. This finding precipitated Lindy's release from Berrimah Prison in Darwin where she was serving a life sentence for the murder of Azaria.

During Lindy Chamberlain's murder trial Crown prosecutor Ian Barker appealed to the jury's common-sense knowledge of dingoes who were not, he said, 'notorious man-eaters'.[14] He capitalised on the dingo jokes that had been circulating since Azaria's disappearance, describing the 'eccentric snow dropping dingoes' of Uluru as they ran about 'resting from time to time with their suitcases and portmanteaus, washing and assorted articles of underwear, stolen from tourists'. He presented Lindy's account of the dingo taking Azaria as far-fetched, unbelievable, 'a transparent lie' that 'would be laughed out of court'.[15]

But those who lived at Uluru – Aboriginal people, rangers, police – and the tourists camping nearby on the night Azaria disappeared accepted Lindy's account. There had been many dingo attacks at the campground and around the Rock in the months and days before

---

10 Bryson 1986, 95
11 Chamberlain 1990, 166
12 Chamberlain, 637 (Barbara Tjikadu quoted)
13 Morling 1987, 241 (Derek Roff quoted)
14 Young 1989, 27 (Ian Barker quoted)
15 Young 1989, 28

Azaria disappeared. On 23 June a dingo grabbed a three-year-old girl, Amanda Cranwell, by the neck and dragged her from a car. Her parents found her lying on the ground with a dingo standing over her, bite marks at the back of her neck, across her left shoulder and on the left side of her face.[16] Ranger Ian Cawood shot Ding, the dingo believed to be responsible for the attack.[17] In July other children had been knocked down and attacked by dingoes, one of whom ran off with a boy's soccer ball in its mouth.[18] Rangers had shot nine dingoes in the two months before Azaria's disappearance because they had become too familiar with people and too daring.[19]

For two years Roff had been writing to the government warning about the dangers of dingoes. About forty lived in dens around the base of the Rock. About twelve or fifteen of them acted as though the campground and buildings in it were part of their territory and their behaviour was becoming increasing bold.[20] Four of them, according to Cawood, or twenty of them, according to Roff, would have the temerity to enter a tent.[21]

On Saturday 16 August a dingo snapped at a hiker and grabbed the cardigan of his fourteen-year-old daughter and the trousers of his ten-year-old son as they walked around the Rock.[22] Late that afternoon a 'sleek, healthy looking dingo'[23] latched onto and shook the elbow of twelve-year-old Catherine West as she sat in a deckchair outside her family's tent to write in her diary. She screamed for her mother. At first the dingo was undeterred, retreating only slowly when Judy, Catherine's mother, advanced angrily.[24] Later Judy West scared off a dingo 'snow-dropping', ripping clothes off a camp clothesline.[25] After dusk in the camping area a dingo bit a nine-year-old boy and stood over

---

16 Dawson 2002, 66
17 Young 1989, 226
18 Bryson 1986, 26
19 Young 1989, 230
20 Bryson 1986, 25
21 Young 1989, 232
22 Morling 1987, 281
23 Chamberlain 1990, 23
24 Bryson 1986, 20–1
25 Bryson 1986, 21

him on the ground. As in some of the other instances, the dingo was slow to move off.[26]

The next night a dingo peered inside a Kombi van as its occupants cooked dinner and stayed while they took two flash photos; one followed a woman as she walked back from the bins; just before 8 p.m. a dingo passed through the edge of a campsite as a mother and daughter washed up their dinner dishes.[27]

Just after 8 p.m. Lindy Chamberlain saw a dingo coming out of the tent where she had put Azaria to bed. She saw the animal's shoulder. Bushes and the shadows of a low post-and-rail fence blocked her view of the lower part of the tent. The dingo looked as though it was having trouble getting through the tent flaps. It swung its head. She thought it had one of her husband Michael's shoes but she couldn't see what was in its mouth. She yelled and it ran in front of their car in the shadow of the fence. Lindy 'dived' straight into the tent, to see what had made Azaria cry.[28] The rugs were scattered. Azaria's four-year-old brother, Reagan, was still asleep, completely enveloped in his sleeping bag. Azaria's bassinet was still warm but empty.

Lindy backed out of the tent, feeling around with her hands in case Azaria was there. She called her husband, Michael: 'My God. My God. The dingo's got my baby',[29] and ran in the direction she'd seen the dingo go. There was a dingo with its back to her, its head turned slightly, in the shadow of their Torana, which was parked alongside the tent. But as she told fellow campers and Senior Constable Frank Morris that night, and as she repeated to Inspector Gilroy the next day, and again to Detective Sergeant Graeme Charlwood six weeks later: 'I couldn't tell you whether it had anything in its mouth or not. My mind refused to accept the thought that it had her in its mouth, although I knew that must be it'.[30]

All the campers searched. Among the many meandering animal tracks on the dune just west of the Chamberlains' tent Murray Haby, a schoolteacher from Tasmania, found fresh-looking tracks that appeared

26   Morling 1987, 281
27   Bryson 1986, 36–7
28   Bryson 1986, 75
29   Bryson 1986, 40 (Lindy Chamberlain quoted)
30   Bryson 1986, 156 (Lindy Chamberlain quoted)

to belong to a big canid. Between them was a furrow, as though it was dragging something. He followed the paw prints higher and found an oval-shaped depression in the sand about the length of a hand and textured, as though by the imprint of a knitted or woven fabric. Haby didn't know if the drop of moisture beside the indentation was blood or saliva. The prints continued over the top of the hill, but petered out as the sand became more compacted. He retraced the tracks back toward the camping ground and emerged near the Chamberlains' tent.[31]

Before the waxing crescent moon set around 11.15 p.m. that night, Haby and Roff followed Nuwe Minyintiri, an Aboriginal tracker, up the dune. Minyintiri lit his way with a firestick made of spinifex. He found two more depressions in the sand, where the 'big feller', as he described the dingo, had rested its bundle. The impression came from something heavier than the usual fauna in the area and, like Haby, Roff thought it had been patterned by some sort of human-made fabric, crêpe or a knit. Touching the perimeter of one of the imprints Minyintiri said, 'Not move anymore'.[32] It was too dark to continue.

The next day Constable Morris, with Inspector Michael Gilroy and Sergeant John Lincoln who had just flown in from Alice Springs, inspected the paw prints that ran alongside the Chamberlains' tent. Morris had found a couple of sets of tracks – he was not sure whether they were dingoes' or dogs' – leading south from the entrance to the tent; turning left, eastwards, at the corner; and travelling between the tent and the Chamberlains' Torana. The police officers saw sprays of blood and what they deemed to be saliva on the tent.[33] Near the corner where Azaria had slept was a heavy, static paw print.

An Aboriginal elder, Nipper Winmarti, spoke to the police on behalf of the group of trackers who had gathered at the campsite at dawn. The group – Winmarti; his wife, Barbara Tjikadu; Daisy Walkabout; Kitty Collins and Nuwe Minyintiri – had followed the prints of a big dingo who was dragging something along the dune and down onto the plain. The dingo ran in the direction of Mutitjulu, the waterhole at the Rock. Like

---

31   Bryson 1986, 53–5
32   Bryson 1986, 57
33   Chamberlain 1990, 58

other dingoes in the area, he used the road, moving into the scrub when vehicles approached.[34] They lost his trail on the road.

When Inspector Gilroy interviewed Lindy that day she told him the dingo 'looked to me a youngish dog'.[35] It didn't look like the mangy dingo that had appeared at their campfire earlier, which Lindy had reprimanded her husband Michael for trying to feed.

When Detective Sergeant Charlwood interviewed Lindy, he thought her story about a dingo staring at her the afternoon before Azaria disappeared was too convenient. Witnesses corroborated her account. She was carrying Azaria, who was awake, unwrapped and looking around, over the rough ground near a site then called the Fertility Cave, about a kilometre west of where Azaria's clothes were eventually found, when she noticed a fit, healthy, reasonably young dingo quietly and intently watching them from a vantage point on top of a boulder a few paces away. The dingo stared at them for at least four minutes.[36] Lindy found it creepy.[37]

In the six weeks between Azaria's disappearance and Charlwood's thirteen-hour interview, recorded without Lindy's knowledge, her impression of the dingo's gender had changed. Originally she thought it was a male but she told Charlwood, 'To my mind, it was a female. Of course, you know, it's supposition. Of course, you probably know more about that than me, by now.'

He responded, 'No. I don't know a great deal about the dingo,' before he asked her to describe again what Azaria was wearing.[38]

In the early 1980s people in the cities and towns of mainstream Australia knew of no precedent for the way Azaria had purportedly died. Like Charlwood, many people knew little about dingoes. At the first inquest into Azaria's disappearance, dingo researcher Alan Newsome, who worked at the Commonwealth Scientific and Industrial Research Organisation (CSIRO) Wildlife Research Division in Canberra, reported that, contrary to what experts had previously

---

34   Bryson 1986, 71
35   Bryson 1986, 76 (Lindy Chamberlain quoted)
36   Bryson 1986, 150
37   Chamberlain 1990, 29
38   Bryson 1986, 159

believed, they had been surprised to discover that dingoes are not loners; they belong to groups.[39]

Minutes before Lindy raised the alert that Azaria was gone, Judy West, camped at the site next to the Chamberlains', heard a growl. It reminded her of the way their farm dogs at home near Esperance in Western Australia growled menacingly to warn each other off when her husband killed a sheep and they were vying for bits of offal.[40] Newsome interpreted the growl Judy West heard as an interaction between two dingoes. He considered that the dingo Lindy saw standing still 'was not the animal which came out of the tent'.[41] The Aboriginal trackers told Lindy there were two dingoes at the tent.[42] Roff thought so too.[43]

Newsome thought that the act of predation of a dingo on a baby was improbable, but possible.[44] Nevertheless, he and Laurie Corbett, his colleague at the CSIRO Wildlife Research Division, thought that the dingoes at Uluru, who were neither tame and biddable nor wild and kept themselves scarce, should be killed and replaced with another generation of dingoes, who could be kept wild.[45]

At the first inquest Nipper Winmarti spoke through a Luritja interpreter, Pamela Harmer, when he told Coroner Denis Barritt about tracking a hungry, thirsty dingo with a distinctive paw print from the dune toward Mutijulu. John Bryson, author of the influential 'true crime' book *Evil Angels*, recounts how Barritt pushed Winmarti and Harmer for responses on the taboo subject of the dingo spirit dreaming. Although Harmer clearly did not want to ask or answer the questions, Barritt persisted. During court recess Harmer went in a rage to the coroner's chambers to tell him the dingo spirit dreaming was a forbidden subject and now Winmarti and his clan had been shamed. According to Bryson, Barritt pursued this line of questioning because, although – at that time – the coroner did not know of any Aboriginal children or babies being taken by dingoes, he thought such an event

---

39 Bryson 1986, 226
40 Bryson 1986, 39
41 Bryson 1986, 227 (Alan Newsome quoted)
42 Bryson 1986, 461
43 Chamberlain 1990, 158
44 Bryson 1986, 228–9
45 Bryson 1986, 230

may be recorded in mythology. Unlike the rangers, police and Aboriginal people at the Rock, who believed that a dingo was capable of abducting an infant, Barritt thought that urban Australians needed convincing.[46] But the coroner extracted the evidence he needed without respecting Indigenous protocols. As novelist and essayist Alexis Wright puts it, the law courts 'want to hear and argue the Aboriginal story from the professional point of view ... in the language of the court'.[47]

Later, at the royal commission into the case, Lionel Perron, father of the then NT attorney-general, Marshall Perron, told Justice Trevor Morling that in 1961, when he was working as an engineer surveying in the Great Sandy Desert, his party camped near some Aboriginal people who speared a semi-domesticated dingo because it had taken and consumed a twelve-month-old infant. Perron's party found the child's partly eaten remains – a mutilated skull and a few bones.[48]

Lindy knew of the pups at the den where Azaria's clothing was found and later acknowledged that 'the dingo was probably returning home with its kill to feed them'.[49] In early 1981, after the first inquest had found death by dingo, the Chamberlains visited Uluru with a film crew to make a television documentary. They filmed the place where the tourists had stumbled upon Azaria's jumpsuit, and the track that led from there to a dingo den, which was still occupied with fresh paw prints around it. The rocks beside the narrow openings on the track were shiny, polished with the oil from dingoes' coats. One of the entrances to the den could only be approached through thick scrub and boulders but a side entrance opened, without hindrance, onto the track leading to the site where Azaria's jumpsuit was discovered.

Lindy wriggled into the hole near the 'main' entrance to the den to take some long-exposure photographs. There she found a cavity in the rock, like 'a very deep baby's bath',[50] full of small bones and skulls of animals, and droppings – she assumed the puppies used it as a toilet area. She knew that Azaria's teeth, formed in her gums,

---

46 Bryson 1986, 217–8
47 Wright 2016, 12
48 Morling 1987, 280–1; Simper 2010; Young 1989, 229
49 Dawson 2002, 74
50 Chamberlain 1990, 175

were harder than her bones and if anything was left of her it would have been her teeth in dingo scats.[51] Her description of this scene in her autobiography focuses on practicalities: the difficulty of accessing the den and balancing the camera; how they might have used a remote-control arm to dig in the area to try to find Azaria's matinee jacket, which at that stage had still not been found. She makes no allusion to how emotionally harrowing this exploration was, how symbolic this strange opposite of rebirth must have been.

The TV documentary was eventually sold to the ABC's *Four Corners*, and police saw the rough footage even though the Chamberlains wanted the footage to be given to their lawyers first. Lindy feared that the puppies' scats, possibly important evidence, would be tampered with and, she writes, her fears were confirmed when they returned to the Rock: the den was no longer occupied, its inhabitants had been killed or moved on; the refuse area was as 'clean as a whistle'.[52]

On 23 July 1983, after Lindy Chamberlain had been jailed, Barbara Tjikadu told (by then) Inspector Graeme Charlwood that the tracks near Azaria's clothes were made by the same dingo whose tracks they had seen at the tent and on the dune the night she disappeared. Constable Morris said that he had not been told of this connection at the time, but it was the first time anyone had taken a statement from Barbara Tjikadu.[53]

In 1986 when Barbara Tjikadu gave evidence to the Morling Royal Commission, Michael Adams, counsel assisting the Crown, asked her how she knew the dingo was carrying a child, not some other prey.

'Because I know if it kills a joey, it will take off with it, carry it,' she answered.[54]

Adams and Commissioner Morling repeatedly asked Tjikadu whether the dingo could have been carrying a joey.[55]

'Was a kangaroo living in the tent?' she answered.[56]

---

51   Chamberlain 1990, 162–3
52   Chamberlain 1990, 175
53   Morling 1987, 247
54   Chamberlain 1990, 635 (Barbara Tjikadu quoted)
55   Chamberlain 1990, 636
56   Chamberlain 1990, 636 (Barbara Tjikadu quoted)

After the attempt to cast doubt on her knowledge of dingoes, Adams tried to discredit Tjikadu's tracking expertise. He asked whether the tracks at the tent could have belonged to a different dingo from the tracks near the clothes.

'If I come to a spot where there's three or four different dingo tracks there,' Tjikadu said, 'they might all be big. I know which is the mother, which is young and so on, and which is old and which is so and so.'[57]

Adams pressed her again to tell him what enabled her to tell the difference between one dingo track and another before Tjikadu's interpreter, Marlene Cousens, intervened.

'I would like to tell you something first before you ask questions like that. When Aboriginal people see tracks, they know who it belongs to, what person went there, because they know the tracks, whereas if all these people got out of the courtroom now and walked barefoot, you can't tell, can you?'

'No,' exclaimed Adams.

The interpreter said, 'Aboriginal people can.'[58]

Tjikadu's ability to read tracks sounds as natural to her as reading a sentence is to someone who has been taught to read letters and words. The court had trouble acknowledging this form of literacy, just as Australian institutions do not recognise Indigenous knowledge of dingoes.

There were two inquests, a murder trial, a royal commission, a pardon, a quashing of convictions and an acquittal. The Chamberlains divorced, each remarried, received a compensation payment and were dissatisfied by the findings of a third inquest in 1995, which left the cause of Azaria's death as open. It was not until 2012 that the Chamberlains' original and consistent claim that their daughter had been taken by a dingo was upheld by the Australian legal system.

The fourth inquest opened in Darwin in February. The Chamberlains' lawyers presented the Northern Territory coroner with evidence of twelve significant dingo attacks that had taken place since 1995, including a fatal attack in 2001 on a child on Fraser Island, known to its Butchulla Aboriginal custodians as K'gari.

---

57   Chamberlain 1990, 637 (Barbara Tjikadu quoted)
58   Chamberlain 1990, 637–8

Nine-year-old Clinton Gage was camping with his family at Waddy Point, on the east coast in the north of the island. Before breakfast on 30 April he and a seven-year-old friend set off up the Binngih track to the Waddy sand blow, a twenty-minute circuit. The boys were not aware that two dingoes were stalking them. According to the seven-year-old's statement in the Queensland Police Service report, the dingoes came from behind. Clinton ran in front of his friend and fell. The friend kept walking with the dingoes following. At some stage a dingo licked his hand. When Clinton's friend stopped briefly, the dingoes returned to where Clinton was lying on the ground and looked as though they began to sniff him.[59] From the footprints left in the sand, a tracker called Brian Little ascertained that one of the dingoes lunged at Clinton from the scrub. He struggled to get up and free himself, walking backwards. When he fell, the dingo struck again and mauled him. Clinton's friend walked with his eyes straight ahead over the big sand dunes and back to camp by another route.[60]

When the two boys had not returned from their walk Clinton's father, Ross, went out with his younger son, six-year-old Dylan, to look for them. One hundred and fifty metres away from camp he found Clinton's body on the track with the dingoes still there. While Ross Gage retrieved Clinton's body a dingo harassed Dylan. Ross Gage kicked the dingo away.[61]

Dr Paul Anderson, the Queensland government medical officer, initially thought Clinton had been attacked by a large number of dingoes because horrendous injuries covered the top of his head, the back of his neck, his chest, his abdomen, his thighs and his groin. He likened the wounds to a shark biting and rotating its head to tear the flesh of its victim. Clinton died from a massive haemorrhage some time after Timothy left and before Ross Gage arrived. The mauling had opened the femoral artery in his thigh, which supplies blood to the leg. According to Anderson, the dingoes were attacking Clinton as a food source.[62]

---

59   Appleby 2015, 136
60   Channel 5, 2001
61   FIDO 2001; Martin 2001a, 2001b
62   Channel 5, 2001

I grew up with the Azaria Chamberlain case: the dingo jokes; the debates, which continue to this day, about whether Lindy murdered her daughter. As Lindy said, 'No one sat on the fence'.[63] For all its prominence, for the huge amount of press and legal attention it garnered, some players and some aspects of the story are elusive. When my mother and I camped at the Rock on a bus tour in September 1980, a few weeks after Azaria disappeared, we saw no dingoes. Since then, the campground has been moved away from the base of the monolith to Yulara, a resort town fifteen kilometres away. Though the Rock itself is seen as timeless, ancient and unchanging, conditions for human traffic around it have changed. In 1980 tourists walked wherever they wanted to. In 2019 the climb was finally closed, in accordance with the wishes of the Rock's traditional Anangu custodians. In 2014 when my family and I cycled unwittingly past the place where Azaria's clothes were found, the track veered away from the base of the Rock around Pulari, a sacred women's site that tourists are not permitted to enter, photograph, video or paint. There is much about the Rock, and dingoes, that is mysterious. To understand them we need other forms of literacy and different ways of seeing, as Aboriginal custodian Kunmanara explains:

The tourist comes here with the camera taking pictures all over. What has he got? Another photo to take home, keep part of Uluru. He should get another lens – see straight inside. Wouldn't see big rock then. He would see that Kuniya [sand python] living right inside there as from the beginning. He might throw away his camera then.[64]

In her report on the inquest into the death of Azaria Chamberlain, Northern Territory Coroner Elizabeth Morris cited some of the fatal attacks by dingoes and dingo crosses on children between 1986 and 2010, including the one on Clinton Gage.[65] On 12 June 2012 she handed down her finding that 'The cause of [Azaria Chamberlain's] death was as the result of being attacked and taken by a dingo'. She addressed Azaria's

63   ACA 2012
64   Parker 2006, 265 (Kunmanara quoted)
65   Morris 2012, 3–4

relatives gathered in the courtroom: Lindy; Michael; one of her brothers, Aidan; and her extended family. 'Please accept my sincere sympathy on the death of your special and loved daughter and sister Azaria.' Her voice cracked on Azaria's name. 'I'm so sorry for your loss. Time does not remove the pain and sadness of the death of a child.'[66]

* * *

Although the fatal attacks on Azaria Chamberlain at Uluru and Clinton Gage on K'gari colour people's perceptions, the ostensible reason people are violent toward dingoes is because they prey on domestic animals used by humans. Legislation around dingoes varies from state to state, and within each state. In Victoria, for example, dingoes are protected but wild dogs are regarded as pests; nevertheless, in some areas, dingoes are not protected on public land within a three-kilometre buffer zone of private land. In New South Wales the term 'wild dog' covers all wild-living canids – dingoes and free-ranging domestic dogs, and their progeny. They are classified as pests under the *Biosecurity Act 2015* (NSW) and land managers are required to eradicate them. In most of Queensland dingoes are legally pests but on K'gari, a World Heritage-listed national park, they are protected.

The names are laden: the public perceives 'dingoes' as native to Australia but 'wild dogs' are regarded as pests, even though the designation 'wild dog' is not accurate. Contrary to earlier studies claiming that most wild-living canids in New South Wales are 'hybrids' or 'feral dogs',[67] (Newsome and Corbett 1985; Stephens et al. 2015) recent genetic research shows that feral dogs are less widespread in eastern New South Wales and dingoes are less affected by hybridisation than previously thought.[68] (Cairns et al. 2020) Scientific names, which attempt to trace dingoes' prehistoric genealogy, also carry human preconceptions and affect how dingoes are treated. *Canis lupus dingo* describes the dingo as a subspecies of the wolf. Like wolves, dingoes live in family groups and breed once a year. *Canis familiaris dingo* denotes

---

66 ABC news 2012
67 Newsome and Corbett 1985; Stephens et al. 2015
68 Cairns et al. 2020

that the dingo is a subspecies of the domestic dog. Dingoes look like dogs and many, including the people I interviewed for this book, call them dogs.[69] But these names have life and death consequences: if dingoes are dogs, once domesticated, they are now feral. As Fiona Probyn-Rapsey points out, 'feral' animals are killable.[70] While the scientific names reflect a dichotomy between wild wolf and tame domestic dog, dingo behaviour does not always fit into European-Australian categories of what constitutes wild and tame. Poet Geoffrey Dutton expresses this paradox when he describes 'two dingoes, so wild they are tame' sniffing 'curiously as puppies, circling closely' in the Everard Ranges in northern South Australia.[71]

As Barbara Tjikadu explained to the Morling Royal Commission, 'A dingo is a dingo.'[72] Tjikadu's statement, based on Aboriginal people's long observation and deep knowledge of dingoes, accords with a third scientific name, which, like the other two, is not without its controversies. *Canis dingo* describes the dingo not as a dog or as a wolf, but as itself, a unique canid, though I do not want to use this term in the way that some conservation biologists and dingo geneticists might, as a designation of dingo genetic 'purity'.

Many Aboriginal languages do not distinguish between dingoes and dogs but they do distinguish between tame dingoes and dogs, and wild-living dingoes and dogs. The Jankuntjara of the Everard Ranges call tame dogs and tame dingoes *papa* and wild dogs and wild dingoes *papa inura*. For the Jankuntjara the distinction between domesticated and wild is also a distinction between inedible and edible. Animals

---

69 While I'm discussing names for dingoes, a note on the naming of people is apposite. Several chapters ('Coolooloi', 'Eurong', 'Let's dance', 'Brothers', 'What they're capable of', 'Traces' and 'Wongari') are based on in-depth interviews with people professionally and personally connected with K'gari's dingoes. The people I interviewed for these chapters have all had the opportunity to read and review the chapters based on their interview. Names of interview participants who did not wish to be identified have been changed, which is indicated at the first mention of their pseudonym in the text. Other people whose identity is protected with a pseudonym include Queensland Parks and Wildlife Service (QPWS) staff, commercial tour operator rangers and others.
70 Probyn-Rapsey 2016
71 Dutton 1967
72 Chamberlain 1990, 635 (Barbara Tjikadu quoted)

that are termed *inura* can be killed but to kill a non-*inura* animal, even accidently, is considered wrong.[73] Similarly the Anbarra people of Arnhem Land, who speak a dialect of Gu-jingarliya, use the same word, *kulakula* or *gulukula*, for dingoes and dogs, but dingoes and wild dogs are called *an-gugurkuja*, which is derived from a verb stem that means 'to be frightened' or 'fearful one' and refers to how dingoes and wild dogs run away from people. Another common Anbarra term for dingoes and wild dogs is *an-mugat*, derived from a noun stem that means 'wild animal or beast' and can refer to a dangerous or solitary man or outlaw, or an unruly woman.[74]

Historic records from New South Wales indicate *mir* is the common root for the word for camp dingoes along the Darling, Murray, Murrumbidgee and Lachlan Rivers, immediately west of the Blue Mountains, on the northern tablelands and along the central coast. In the late 1800s and early 1900s in Wollongong and Botany Bay *mirrigung* and *mirigung* were the words recorded for tame canids; at Forbes and Condobolin *mirrie* and *mirree*; *mirrigan* at Narrandera and *mirree* and *meeree* at Moree. Place names such as Merri-Merrigal (Bourke), Mirrie (Dubbo), Merri Merri (Wellington), Merry (Lachlan River) and Mittagong (from the 'Moneroo' or Monaro) derive from this root.[75]

The most prevalent root for wild canid in New South Wales is *(y)urig, -ag-*, which was recorded as *yuggi* at the Namoi and Barwon Rivers, *yukey* at the Macquarie and Castlereagh Rivers, *euchie* and *yukey* at Dubbo and *joogoong* at Botany Bay. From this root come place names such as Ureggin (Casino), Touragon (Ballina) and Youroogin (Murwillumbah). *Waregal*, another root for wild dingo, was recorded around Port Jackson.[76]

In 1788, when the British arrived on the land of the Gadigal people of the Eora nation at Warrane (the place that they named Sydney Cove), some dingoes were wild, living in their own cooperative, communicative social groups, and some were tame, living with Aboriginal people.[77]

---

73 Hamilton 1972, 287, 290
74 Meehan et al. 1999, 91–2
75 Ryan 1964, 112–5
76 Ryan 1964, 117
77 Breckwoldt 1988, 56–78

Aboriginal people kept animals including birds, possums and young wallabies but dingoes are the only ones that have been found buried in the same manner as people.[78] After European settlement, Aboriginal people embraced European domestic dogs[79] and now camp dogs are a firm feature of some Indigenous communities. Some aspects of contemporary human–canid relationships may be similar to historical relationships between people and dingoes; some are no doubt different. Accounts by eighteenth-, nineteenth- and twentieth-century European mariners, explorers, surveyors, geologists, missionaries, administrators and anthropologists record a variety of relationships between Aboriginal people and dingoes, which reflect the diversity of Indigenous societies: some people eat dingoes;[80] some prefer not to;[81] some breastfeed them as they nurse their children;[82] in some places dingoes help in the hunt;[83] in some communities dingoes and dogs accompany women and children on their foraging expeditions but are actively discouraged from accompanying men on their hunts;[84] dingoes smell and dig out water from the sand of stream beds, which helps people and birds;[85] canids provide warmth and companionship.[86] These companion canids are individuals with names: Yelabeli and Happy's dog;[87] Pinaltju (Listener), Panari (Digger), Tiyu (Sparks), Papi (Puppy), Lasi (Lassy), Nipa (Nipper) and Tjaputi (Dirty Mouth).[88]

---

78 Koungoulos 2020
79 Jones 1970
80 Tindale 1974, 36, 109; Meggitt 1965, 14; Hamilton 1972, 288–90; Smyth 1972, 148; Giles 1986, 20; Breckwoldt 1988, 65
81 Meggitt 1965, 14; Gould 1969, 261; Meehan et al. 1999, 98; Mountford 1981, 184
82 Mitchell 1965, 347; Berndt and Berndt 1942, 162; Philip 2017b, 93; Dixon and Huxley 1985, 166
83 Meggitt 1965, 19; Chewings 1936, 32; Smyth 1972, 147, 190; Bates 1985, 247
84 Basedow 1925, 119; Gould 1969, 263; Hamilton 1972, 291; Kolig 1978, 91; Meehan et al. 1999, 102
85 Tindale 1974, 120
86 Meehan et al. 1999, 97; Meggitt 1965, 15; Hamilton 1972, 292–4; Tindale 1974, 109
87 Meehan et al. 1999, 97
88 Hamilton 1972, 294

The first printed use of the word 'dingo' in English is attributed to Watkin Tench, a marine who came to Sydney with the First Fleet and published an account of New South Wales: 'the only domestic animal they [Aboriginal people] have is the dog, which in their language is called Dingo, and a good deal resembles the fox dog of England'.[89] The word 'dingo' originated at Port Jackson and there has been speculation that Tench and other early settlers misunderstood *tingo*, a word for tame, which was corrupted into 'dingo'.[90]

Newton Fowell, who in 1788 was a midshipman on the *Sirius*, reports in a letter home to his family in Devon that Aboriginal people around Sydney Cove 'have fish bones claws of Birds or a Dogs tail tied to their hair & gumed that it might not come off'[91] and that:

> They have a number of Dogs belonging to them which they call Tingo, they do not bark like our Dogs but howl, the Govonor has one of them that he intends Sending home in one of the Transports, they are the Wolf Dog – are the Colour of a fox & have a brush tail at first would eat nothing but fish that being his constant food.[92]

Writing in the 1830s, after expeditions into the interior of eastern Australia, surveyor Thomas Mitchell describes how:

> The Australian natives evince great humanity in their behaviour to these [native] dogs. In the interior, we saw few natives who were not followed by some of these animals, although they did not appear of much use to them. The women not unfrequently suckle the young pups, and so bring them up, but these are always miserably thin, so that we knew a native's dog from a wild one by the starved appearance of the follower of man.[93]

---

89 Tench 1789, ch. 11
90 Breckwoldt 1988, 72
91 Fowell 1788, 21
92 Fowell 1788, 23
93 Mitchell 1965, 347

Anthropologists Roland and Catherine Berndt record Aboriginal women breastfeeding pups and other intimacies at Ooldea in western South Australia in the early 1940s: 'Sometimes a pup may be suckled by a woman whose child has recently died, while they are often played with by all the camp. Idly while talking, a man may sooth a tired dog by fingering its penis.'[94] In 1916 on Mornington Island, John William Bleakley, Chief Protector of Aborigines, photographed a Lardil woman breastfeeding two week-old pups.[95] In the 1920s on the Edward River on the western side of the Cape York Peninsula, zoologist and anthropologist Donald Thomson records how a Koko Dai-yuri man, Tjamindjinyu (Tommy), told him

> that when the puppies are captured very young they are fed on meat hammered to pulp, and if too young for that, are fed at the breast ('tjo 'tjo) by the women. His own wife, Chako, fed one of the dogs he still has in this fashion. When rearing it – as Tommy put it, the child had one side, the puppy the other.[96]

The interspecies bonds created in part by such nurturing were profound and did not always make sense to European observers.

Thomas Mitchell could not see the use of dingoes to their Aboriginal companions or, for that matter, the use of people to their dingo companions but unwittingly he records other less tangible aspects of the relationship. On 1 May 1836, out along the Lachlan River in the vicinity of what is now Booligal, Mitchell's party broke camp but:

> [j]ust as the party was leaving the ground a noise was heard in the rear, and two shots were fired before I could hasten to the spot. These I found had been inconsiderately fired by Jones our shepherd at a native dog belonging to our new guide and which

---

94 Berndt and Berndt 1942, 162
95 Philip 2017b, 93. As Justine Philip (2017b) argues, interspecies 'wet-nursing' was the only way of keeping mammalian young alive before modern technology enabled humans to artificially feed human and animal infants, and probably played an important role in animal domestication.
96 Dixon and Huxley 1985, 166 (Donald Thomson quoted)

had attacked the sheep. This circumstance was rather unfortunate, for our guide soon after fell behind, alleging to the party that he was ill.[97]

Mitchell was pragmatic, more concerned about how his party was going to find water that day without their guide than the demise of a dingo, but in two sentences, a tangential aside to his main task of exploration, he reveals an attachment that, baffling and perhaps unbelievable to the newcomers, can transmit itself across space with bodily ramifications: Mitchell's guide was not physically harmed but the European shepherd's killing of his dingo made him sick. If, as mining engineer and ethnologist Robert Brough Smyth writes, 'nothing more offends a black man than to speak harshly to his dogs, or to depreciate them: and if any one gave a black man's dog a blow, he would incur bitter enmity',[98] the magnitude of the offence of the shooting would have been beyond comprehension on the Europeans' part and the motivation for the shooting beyond comprehension on the Indigenous people's part.

The Butchulla language, like many other Aboriginal languages, does not distinguish dingoes from dogs, but does distinguish tame dingoes and dogs, which are called *wadja*,[99] or *wat'dha*[100] from wild-living dingoes and dogs, called *wangari*[101] or *wongari*.[102] The differences between these two canids are described from a Butchulla perspective in a few informative paragraphs in a 2017 QPWS dingo safety and information guide: 'Wat'dha were our companions – always part of us. They helped us hunt and track, and protected us from bad spirits and the Wongari. Wongari have been and always should be wild. They are a natural and important part of the ecosystem on K'gari'.[103] The guide explains the disappearance of the wat'dha companion dingoes as a consequence of colonial history and Aboriginal dispossession: 'When the last of our people were taken off the island, all of the dingoes

97   Mitchell 1965, 59
98   Smyth 1972, 147
99   Bell and Seed 1994, 135
100 QPWS 2017, 2
101 Bell and Seed 1994, 136
102 QPWS 2017, 2
103 QPWS 2017, 2

became wild, but we, the Butchulla, are still all strongly connected in our hearts, minds and spirits.' The guide asks visitors to 'respect Butchulla lore' because what's good for the land comes first. It continues, 'K'gari is Wongari Djaa (Country), and provides everything they need. They are curious, but need you to keep your distance. So please, don't feed Wongari.'[104]

According to this brochure, the close canid–human bonds that have existed in Australia for at least a millennium – and probably much longer – have been broken on K'gari. If I am interpreting the brochure, such an ephemeral form for such devastating history, correctly, it is saying that a couple of hundred years of colonisation have erased K'gari's wat'dha. But what do contemporary dingoes make of these changes? How are they adapting? What happens if some of them seek a different sort of relationship with humans?

I was scared and thrilled when I met a dingo on the beach during my first research trip to K'gari. Our encounter lasted just over one relatively uneventful minute. Since, I have pored over that reverberating moment, trying to remember it accurately. Here is my story of the dingo I called Bold and how I came to know him.

---

104 QPWS 2017, 2

2
Strange familiar country

The ground came first. After that her feet and eyes and mouth.
Stephen Daisley, *Coming Rain*, 2015

'Welcome to the Queensland government. If your call is in relation to a medical emergency, please hang up and dial triple zero.

'If your call is in relation to a bite or scratch from flying fox or other bat, press eight. For all wildlife enquiries including crocodile and cassowary incidents, press one. For reporting or making enquiries regarding pollution, fish kill or unauthorised interference with a cultural heritage place, press two. To report illegal activities occurring in Moreton Bay Marine Park, press three. For enquiries related to licences for activities regarding native plants and animals, environmentally relevant activities including resource activities, contaminated land and compliance assessment, press four. For camping permits, organised activities in a national park or forest, connect-with-nature activities in south-east Queensland, press five. For all other enquiries, press zero. To repeat this menu, press the star button.'

Which I did.

Sitting at my desk in our study in a suburb of Sydney, I had dialled the toll-free number for Queensland Parks and Wildlife Service (QPWS) to request an interview with rangers about dingoes on K'gari, also known as Fraser Island. I wished to talk with QPWS rangers about relationships between dingoes and people on K'gari, to learn how people's perceptions of dingoes and emotional responses to them affect how they are managed.

The voice that related the list of callers' possible concerns was a man's voice with a broad Australian accent. It said 'camping' with closed lips so it sounded more like 'kemping'. The 'o' in 'cassowary' was clipped and the 'a' sounded like 'e' so the last two syllables rhymed with 'ferry' not 'wary', which is, I guess, the correct way to pronounce it. The voice enunciated 'fish kill' very clearly, and swooped into the 'l' of 'licences' with an emphasis that reminded me of my father when he was drunk.

Unlike the anodyne tones of a Siri or a Google voice, this voice belonged in an outdoor world where fish are killed and crocodiles and cassowaries have 'incidents' with people. This voice may require a person's compliance regarding pollution and contaminated land. This is the voice that may give a person state-approved licence to do things with native plants and animals. This voice survives in a country where, as Northern Territory Coroner Elizabeth Morris pointed out in her report on the 2012 inquest into the death of baby Azaria Chamberlain, between July 2000 and November 2010 there were 254 human deaths in Australia caused by animals taking or attacking people.[1]

After the call, I explained in an email to Senior Ranger Ben Steep,[2] whose name I was given by the person I spoke to at the end of the recorded message, that I wanted to learn how people's perceptions of dingoes and emotional responses to them affect how they are managed. I was planning to travel to the Fraser Coast in May 2015 to attend a public forum about dingoes and I hoped to interview some QPWS rangers while I was there. I received a prompt courtesy email from another ranger, Senior Conservation Officer Naomi Stapleton,[3] advising me that she was seeking advice regarding Parks' involvement from line management. Her immediate manager was on leave until the following week and she would discuss my request with him on his return.

A couple of years before, when I started thinking about dingoes, I had read about dingo researcher and photographer Jennifer Parkhurst.

---

1   Morris 2012, 10
2   A pseudonym has been used here and subsequently.
3   A pseudonym has been used here and subsequently.

She spent more than six years observing dingoes on K'gari before she was prosecuted by Queensland's Department of Environment and Resource Management (DERM) for interfering with a natural resource and for feeding dingoes. QPWS, the organisation that manages most of K'gari, is part of DERM, which is now called the Department of Environment and Heritage Protection. Before her prosecution Parkhurst had been critical of how QPWS were managing dingoes on the island. The Queensland government's heavy-handed treatment of Parkhurst intrigued and scared me. She claimed that she fed a pack of dingoes because they were going to die of starvation; she was on the dingoes' side, she said. Her photographs reveal non-captive dingoes in their most intimate moments: a pregnant mother prepares her den for the birth of her pups; adults regurgitate food for pups; dingoes play and hunt and look fierce and relax.[4] Other books and resources about dingoes come nowhere close to this level of involvement.

I wondered whether the actions of DERM and QPWS against Parkhurst could be viewed as yet another form of state violence against dingoes. I contacted Parkhurst because I was interested in her story, and I asked her whether I could interview her. Later, during our interview, she told me DERM prosecuted her because they had to, as she put it, 'shut me up' and 'get rid of me'.[5]

What happened to Parkhurst also seemed to me, observing from New South Wales, a continuation of the authoritarian and unrestrained use of power that had made Queensland notorious during the 1970s and 1980s when Joh Bjelke-Petersen was premier and corruption rife. In the late seventies, with teenage superciliousness, my elder brother Bruce and I teased our cousins, visiting from Toowong, about Joh and the gerrymandering, police brutality, graft and gifts that were part of public life in their home state. They stood with their backs to the piano, which one of them had just been playing with an innate ear and great skill, and opened their mouths as though they were laughing but their eyes were not laughing. They were too polite to point out that our own New South Wales premiers Robert Askin and Neville Wran were both

---

4   Parkhurst 2010
5   Unless specified otherwise, subsequent direct quotes from Parkhurst are from this interview (Parkhurst 2015b).

alleged to have links to organised crime. When he was in his cups my father ranted about the lies Wran told and the connections between the NSW police, judiciary, criminals and politicians. My eldest brother, Roderick, had grimy newspaper cuttings about Sydney underworld boss Abe Saffron stuck to his bedroom wall. I grew up believing corruption was part of life in Sydney but it lay under the surfaces, unseen by many, and it had a permissive quality – relating to drugs, prostitution and people buying their way out of prison. It was not blatant, repressive and overtly political like corruption was in Queensland. New South Wales corruption, we believed without thinking, was how corruption should be.

QPWS's prosecution of Parkhurst galvanised other critics of dingo management on K'gari and publicised the work of Save Fraser Island Dingoes (SFID), a group formed in 2009 by people who were concerned about whether there was adequate food available for dingoes on the island and whether there were enough dingoes there to form a sustainable population. They were vocal in their opposition to the QPWS practices of killing problem dingoes; deterring dingoes from humans by shooting them with pellets from slingshots (known as hazing); and ear tagging them, which, Parkhurst and SFID claimed, impeded their ability to hunt.

Parkhurst agreed to let me interview her at her house in Rainbow Beach on the Cooloola coast. So in May 2015 I travelled to Hervey Bay, which is about 120 kilometres north of Rainbow Beach, to attend the forum about dingoes, to interview Parkhurst and to visit K'gari. Organised by SFID, the forum, *The dingo – friend or foe?*,[6] was designed to raise awareness and promote discussion about dingoes in Queensland. Keen to meet prominent dingo advocates and listen to local pastoralists, I flew in to Hervey Bay the day before.

It was late morning when my plane descended. I looked down. Wind whipped up whitecaps in the Great Sandy Strait. *Shipwreck*, I thought. *Plane crash.* We had flown over K'gari – wooded, sand-tracked, beach-fringed, its folds and undulations hiding dingoes, its creeks winding their way to the coast. It is a big island, the biggest sand island in the world and Queensland's largest island – 124 kilometres long and

---

6   SFID 2015

between five and twenty-three kilometres wide. Its north-easternmost point, Sandy Cape, is ninety-six kilometres from the mainland coast and its southern tip, Hook Point, is just over a kilometre from Inskip Point, near Rainbow Beach, on the mainland. From the air K'gari's east coast, a long straight line of beach and breakers, is distinctive. In the forested interior an amoeba-shaped lake shone up at us. It might have been the lake that Amanda, Doug[7] and I camped at in the late 1990s.

My friend Amanda, and Doug, her boyfriend at the time, worked at Kingfisher Bay Resort Village, a new ecotourism resort on the west coast of the island. When we weren't out camping I slept in Amanda's staff donga, a simple rectangular demountable, and she moved into Doug's donga. Doug, who had worked on the island for a long time, took us around on his chook-chaser and in a borrowed Mini Moke to places I would never have been able to get to without him. For a couple of days we stayed at his illicit humpy in the bush. No-go areas were not no-go to Doug. When a fence post blocked a track he simply lifted it out of the sand.

We camped by exquisite lakes alive with the noise of frogs and visited the west coast looking for where the Bogimbah Creek Aboriginal reserve and mission had been. The reserve was started in early 1897 when, in response to complaints by the white residents of Maryborough that young Indigenous women were spreading venereal disease to white men, Archibald Meston, who was Protector of Aborigines for southern Queensland from 1897 to 1904, 'mustered' (his word) Aboriginal people in Maryborough – thirty-three men and youths, and eighteen women and girls – and took them to live at an old quarantine station on the island at White Cliffs, which was known to them as Balarrgan. A couple of months later, after the Aboriginal residents of Balarrgan chased and attacked a Good Friday excursion of white youths from Maryborough, the reserve was moved north to the less hospitable location of Bogimbah Creek. White Cliffs, a popular pleasure spot for white settlers, had been gazetted as a recreational reserve.[8]

After the Anglican Church took over the Bogimbah Creek reserve in 1900 Meston washed his hands of responsibility for the ill health of

---

7   Pseudonyms have been used here and subsequently.
8   Evans and Walker 1977, 74–82

the people there, many of whom had been brought from other parts of the state as he enacted his plans to make Fraser Island a station for 'blacks over whom firm control is a stern necessity'.[9] Rations under Meston had been calculated to supplement hunting and fishing. It was impossible to grow vegetables in the sand, fish were scarce and the church prohibited Aboriginal traditions, so people lived on a monotonous diet of a thin maize porridge called hominy; there was no milk, meat or vegetables.

Before the mission was disbanded in 1904 and most of its surviving inmates were sent to other parts of Queensland, the people incarcerated at Bogimbah suffered from malnutrition, measles, mumps, syphilis, bronchial and chest complaints, including tuberculosis, and what was described as an addiction to eating earth, especially white clay (dulong) from Little Woody Island (Walangoora).[10] The practice of eating earth or soil-like substrates such as clay or chalk is known as pica, or more specifically geophagia, and has been recorded in many places since ancient times. It is not unusual for geophagists to be very discerning about which earth they crave.[11] The people at Bogimbah prepared dulong, and other substances, by drying it in the fire; by putting it in water, which gave the water a milky appearance before they drank it; or by biting pieces off a lump.[12] When the children were prohibited from eating dulong, they ate sand, charcoal, brick, chalk, shell and slate pencil instead. Child mortality was high. Nearly all the women ate earth too.

When Dr Davidson of Rockhampton and Dr Penny of Maryborough visited Bogimbah they diagnosed ancylostomiasis or hookworm disease. Its symptoms were 'an enlarged liver, listless eyes and generally sleepy demeanour, wasting and irritability, weakness, shortness of breath and palpitation of the heart upon exertion'. In a report dated 21 February 1901, Dr Penny wrote:

> once afflicted with this form of disease, patients do not confine themselves to earth ... but eat ashes, charcoal etc. ... weakening

---

9   Evans and Walker 1977, 83 (Archibald Meston quoted)
10   Evans and Walker 1977, 87
11   Young 2012, 5
12   Evans and Walker 1977, 88

symptoms are followed by dropsical effusions, diarrhoea, dysentery and death if left untreated.[13]

Ancylostomiasis occurs as a result of severe infestations of hookworm in the small intestine. When the hookworms attach themselves to the wall of their host's intestine to suck blood, their host becomes anaemic. The female hookworms lay eggs that are excreted in faeces and when people walk barefoot over contaminated soil or sand, larvae can penetrate their skin. The worms travel through the bloodstream to the lungs and migrate from the lungs up the respiratory tree to the mouth. They are swallowed and carried down to the intestine where they attach themselves to the intestinal wall. They suck the host's blood, the females lay eggs and the life cycle starts again. The human-infecting species of hookworm that caused people's anaemia at Bogimbah were probably *Ancylostoma duodenale* and possibly *Necator americanus*.[14]

Could dingoes be implicated in the anaemia of the residents of Bogimbah, I wondered. Dingoes are killed with such alacrity, would even raising this question have bad consequences for them?

Canids can have hookworms, a vet and expert in gastrointestinal parasites told me, but canid hookworms (*Ancylostoma caninum*) rarely develop to maturity in humans' intestines. It is possible that hookworms carried by dingoes had effects on the health of the people at Bogimbah, especially if their relationships were close, but the impact was more likely to be dermatitis, or 'ground itch', from contact with infective larvae.[15] It is unlikely that the people at Bogimbah contracted hookworms from eating earth because hookworm larvae do not survive well in clay and, if they do, they are usually killed by the heating and drying process.[16]

Much colonial writing from many parts of the world – Africa, the Caribbean, South America and South East Asia – condemns Indigenous peoples' consumption of earth as the behaviour of inferior and even depraved people. These preconceptions occlude clear-eyed appraisals of

---

13 Evans and Walker 1977, 87 (Dr Penny quoted)
14 Kopp 2018
15 Kopp 2018
16 Young 2012, 63–4

pica, the motivations of the people who engage in it and its consequences.[17] Perhaps the people at Bogimbah ate earth because of a shortage of food. Perhaps they sought nutrients they were not getting from their diet. Perhaps eating clay was a way of alleviating nausea. But, like ancylostomiasis or hookworm disease, pica is associated with anaemia – even where hookworm infections are rare.[18] Ingesting clay inhibits iron absorption so pica can contribute to micronutrient deficiencies rather than alleviating them.[19] On the other hand, clay has extraordinary binding capacities and is very good at removing unwanted particles.[20] Eating earth may have been a way for people to protect themselves from toxins because clay particles fortify the mucosal lining of the intestine from being eroded by acidic foods, thus reducing the permeability of the gut wall and reducing the chance of food and waterborne pathogens entering the bloodstream.[21] Clay particles may also be able to deactivate toxins by 'grabbing them before they can be digested'[22] and adsorbing them, or binding them to their surfaces. So eating earth, the behaviour that matrons and doctors and reservation and mission managers perceived as self-destructive and dangerous, may have actually been motivated by a deep-seated, gut-wrenching will to survive.

In 1905, after the mission closed, Bogimbah became a timber camp and the 70 graves in the mission's cemetery were obliterated. In the same year, Meston wrote a report on the island's resources and its tourist potential, noting that, '[i]f the dingoes, which are *very numerous*, were exterminated, the island would be an ideal spot for the preservation of all varieties of our native fauna'.[23]

If memory could be exterminated, too, we could all enjoy what the Kingfisher Bay Resort Village's website calls 'the island's untamed wilderness and rugged, natural charm' unimpeded. In 1998 when Amanda, Doug and I tried to find Bogimbah there was nothing to mark

17 Kopp 2018, 70–3
18 Kopp 2018, 114
19 Kopp 2018, 115–16
20 Kopp 2018, 35
21 Kopp 2018, 121–2
22 Kopp 2018, 123
23 Meston 1905, 5, italics in original

the site. I took a photograph of dark grey logs lying in a horizontal mass with a row of grey-black vertical pylons of various heights – short, tall, short, tall, short, tall, short – beside them. In the picture the wood is dense, solid, a contrast to the pewter sea and the watery, pale grey sky. The only colour is the dark green leaves of young mangroves as they sprout up through the logs on the sand. I took photos of swamp lilies and melaleuca; of the expressive whorls and crevices in a piece of driftwood worn smooth by sand, water and wind; and of the tangled buttress roots of mangroves with their thin trunks dressed in lichen and their slight branches reaching out to each other as if they were dancing.

I remember mosquitoes, little shells among small mounds of sand pellets on a stretch of flat, hard-sanded beach, a lank stillness in the air and a sense of desolation that could have come from my friends because they knew much more about the history of the place than I did. They were not certain we were in the right spot. Bogimbah does not appear on my 2014, ninth edition, Hema waterproof four-wheel-drive explorer map of the island, though the creek is marked, flowing out to the strait near Little Woody Island.

White Cliffs is now the site of the Kingfisher Bay Resort Village where, in 1998, I joined tours, when space was available, and listened to Amanda's colleagues talk about this unique World Heritage-listed ecological wonderland to visitors from the city and abroad. The rangers displayed their expertise manoeuvring ungainly four-wheel-drive buses along the island's soft-sanded tracks and enjoyed telling us in sympathetic and melodramatic tones about the sex life of the yellow-footed antechinus, a marsupial whose vigorous twelve-hour mating sessions cause the male's immune system to collapse so he dies.

The people I met who worked at the resort loved the island and accepted the structures of authority that purported to keep it in its so-called pristine state. But, despite talk of paradise, they did not believe in religious stories of paradise and human origins. They believed in evolution, survival of the fittest. We discussed how humans fit into the natural world. I thought humans were different from animals; humans formed the notion that we were made in the image of God; humans talk about humanity, ethics, ideals. There seemed to me to be plenty of evidence that we are not the same as other animals: institutional religion, industrialisation. Some of the rangers argued that humans are just another species; if we humans dominate our environment it is because we are evolutionarily adapted to do so.

I had taken a draft of a novel I was writing with me and one night I checked into a room in the resort to work on it – as if I could resolve the problems with that manuscript in one night! I was having trouble with the central section of the story, which involved slaves, a princess, a sea voyage and a ghost. Because she had signed an Act that emancipated the slaves in her empire, the princess had been forced to leave the southern-hemisphere country her grandfather had settled, and which she regarded as home, and return to the European country her family had come from. The princess's grandfather, an amoral, opportunistic, venal character, was haunting the ship on which she travelled. I was frightened of this ghost and his power. I was afraid of that sea voyage, too, afraid of summoning up ghosts and of trespassing on matters I had no authority to write of.

In the middle of that night a storm blew up. An almighty wind, an unremitting and malevolent gale, caught the resort in its grasp and shook everything and all of us. It threw itself at the windows and came blasting in through the door, which opened. No person had opened the door. It was the force of the wind, the angry spirits. A warning? I never did finish that manuscript.

* * *

Seventeen years later, driving a circuitous route from Hervey Bay Airport to the White Crest Luxury Apartments at Torquay in my uncomfortable rental car, the streetscapes' neatness was a contrast to the havoc of that storm and the meandering shapes of my aerial view of the island.

Hervey Bay is called a city but it is really a locality, an urban area tucked into a curved stretch of north-facing coast that is sheltered from the open sea by K'gari. Along the coast, from east to west, the towns Urangan, Torquay, Scarness, Pialba and Point Vernon have joined up. Hervey Bay Airport is close to town and River Heads, the departure point for one of the barges to K'gari, is a fifteen-minute drive south, so Hervey Bay has become a convenient stepping-off point to the island.

Very short emerald-green grass surrounded single-storey houses arranged in neat oblongs with Australian suburban sparseness. None of the figs, eucalypts, casuarinas or melaleucas, which grow to gigantic proportions on the shorefront in this astonishingly fecund Queensland sand, grew around the houses. Their yards were flat with space for vehicles and boats and other machines and toys. Well-maintained fences. Well-maintained verges.

My spacious apartment had white tiled floors and was furnished with bulky leather lounges and a glass-topped dining table with hard chairs. It was at the back of the building and faced south over the car park and tennis court, not north over the bay. I arranged my things – my mega-boom speaker, my iPad with reading and music, print-outs of scientific articles about dingoes, books, candles, food – and went for a walk.

The weather was colder than I thought it would be and drizzly. No one wore a raincoat or carried an umbrella along the Esplanade. Near Pialba the tree trunks between the path and the shore were pale green with lichen, which was thicker on their southern side. I enjoyed the lichen-scented air redolent of old books, warm stone, breezy weather and sea spray. But the smell of lichen also reminds me of worlds beyond human words and associations, of miraculous processes of absorption and tenacity. British author Helen Macdonald, in her memoir *H is for Hawk*, writes that lichen can survive 'just about anything the world throws at it. It is patience made manifest.'[24] Perhaps that's why, when I encounter lichen, I put my nose close to its soft-firm, faded blue-green fractal shape and inhale. I need patience.

If lichen smells of air and aeons, the flying fox colony at the mouth of Tooan Tooan Creek smelt of earth and now – pungent, animal, the sort of smell people complain about. Flying foxes clustered in trees like heavy fruit, hanging upside down with their wings around them. I couldn't see how they attached themselves. They were in all different kinds of trees – the main thing seemed to be that they chose to hang in a tree with other flying foxes. One tree would have no flying foxes in it and the tree next to it would be jam-packed with little sacks of bats.

Life is so boring in a resort town in the off-season that I was heckled twice from passing cars. Hervey Bay was up for sale. Vacant lots

---

24   Macdonald 2014, 11

were advertised with signs announcing 'Development site approved. High density. Shopfronts'. Next door to the cleared lots gracious old Queenslanders, surrounded by hibiscus, oleander and palms, looked naked, quivering, on their exposed side. Six-storey, block-wide apartments monstered modest 1950s vernacular-style flats, shops and hotels. The easterly had dropped. The tide was coming in. A faint rainbow appeared over where K'gari had disappeared into the clouds.

*　*　*

Dingoes on K'gari have it better than dingoes in most other parts of Queensland, where they are a declared Class 2 pest animal under the *Land Protection (Pest and Stock Route Management) Act 2002* and land managers are obliged to 'control' or kill them. On K'gari they are protected as a native species under the *Nature Conservation Act 1992* (sections 17 and 62) – as they are meant to be protected in other national parks in mainland Queensland. But, as ecologist Arian Wallach pointed out at the *Dingo – friend or foe?* forum,[25] dingoes are baited in programs designed to eradicate foxes, cats and wild dogs in national parks – the very places where they are meant to be safe.

Under low lights in a big room at the Hervey Bay Community Centre in Pialba, Arian Wallach looked small and delicate standing in front of a PowerPoint projection of a cane toad (an animal almost universally reviled in Australia for being an invasive pest), telling the audience that we should stop killing for conservation. She talked about the high economic, ethical, social and environmental costs and risks of killing. She asked us to envisage an ecosystem with all of our natives and our introduced 'feral' pests: bilbies and wallabies, cane toads and cats. She explained how native animals can co-evolve with cane toads.

She also explained how important large predators are to ecosystems. Worldwide, she said, the ecosystems that are working are those that have large predators. But big predators are persecuted and hunted globally. Australia's large predator, the dingo, has no enforced legal protection anywhere. National parks, she said, were some of the most dangerous places for dingoes. She cited studies that demonstrate

---

25　SFID 2015

that where foxes are common, dingoes are scarce; that marsupials persist where dingoes persist; that when predator control stops, biodiversity recovers.

Wallach described how dingoes are social animals who practise family planning. When lethal control disrupts dingo pack structure, she said, they breed faster. Dingo populations stabilise when persecution stops. She argued that stock losses to predators decrease and costs decrease when farmers stop using lethal control. She asked us to regard the dingo as a native – it was here when Europeans arrived to settle in 1788, even though it is not as 'native' as the kangaroo. Canids all hybridise, she said. Wolves, jackals, dingoes, coyotes all hybridise. Wild dogs elicit hatred in Australia, she said, but these 'wild dogs' are dingoes.

At afternoon tea Wallach's partner, Adam O'Neill, also a dingo researcher, said, 'I think she went a bit far with the cane toad.'

Retired farmer Harry Jamieson described how he never baited or trapped dingoes on his beef-fattening property at Tiaro. He never shot kangaroos and wallabies either – they were for the dingoes. He never lost a calf to a dingo. He had witnessed dingoes hunting kangaroos and he had rescued his dauntless cattle dog from dingoes. He used to be able to get within thirty to fifty metres of a particular 'quiet' female dingo – his wife called her his pet – but she always ran away when he got off his horse or stopped the vehicle. 'We've got to live with them,' he said.

With the laconic authority of someone whose family had been interacting with dingoes since 1852, Lindsay 'Butch' Titmarsh explained how food, water and habitat govern animal numbers. He reckoned there were now two or three or four times more kangaroos on Tandora, his 4400-hectare property at the junction of the Mary and the Susan Rivers, than 150 years ago because they have open flats, which they didn't have before European settlement. 'Roos everywhere,' he said. 'When it becomes dry weather you get a few roos, you knock a few over.'

Knock a few over. I could bottle his voice.

Lindsay Titmarsh's family passed down stories of seeing one or two, or three or four 'dogs'. To see six was extraordinary. 'We never see big mobs, they just work in two or three and they can catch a roo pretty easy. We've shot 'em, poisoned 'em, trapped 'em for years. My father's theory was if you've got roos, the dogs are happy. If you see a dog every two months you're doing well,' he said. 'You don't see 'em very often.'

'But the middle of last year, we saw five in a mob. Thought that's a few too many, so we go and sort them out. We have shot sixteen dingoes in the last, probably, nine months. Unheard of at Tandora. And there's still plenty there. Are they dogs coming in from drought country?' He was trying to figure it out. 'Do they come to their neighbour's territory and the neighbour comes to his neighbour's and they end up here? You would not think a dog would come from Charleville straight to Tandora.'

Dairy farmer and Fraser Coast councillor James Hansen also reported a change in dingo numbers and behaviour on his property on the Boompa Creek. Ten years ago, he said, a lot more dogs turned up and they were brazen. The odd calf got 'nibbled' and a few lost their ears and tails. Then packs of about five dogs started pulling down the weaner heifers at night. He explained the financial value of the weaner heifers – $1000 to replace and worth about $10,000 in income over their lifetime. Then the dogs started pulling down calves in the daytime. They started eating calves as they were being born. He could get twenty feet (six metres) from them and they didn't run away.

'I used to go home,' he said, 'grab the .223, go back and bowl the dogs and shoot the poor bloody calf. I don't know where these dogs came from. They were never like that once. There were rumours going around they were released from Fraser Island but I always found that hard to believe.' Until he saw two dogs wearing tracking collars. 'I don't think they lasted long,' he said, 'because one of the neighbours got 'em.'

After the forum, one of the organisers, Karin Kilpatrick from SFID, told me about the stories of rangers dumping Fraser Island dingoes on the mainland. Why would they do that? I didn't understand it. Even if they weren't killed, Kilpatrick intimated, they would die on the mainland. 'They're beach dingoes,' she said, 'not inland dingoes.' I found it hard to believe that QPWS would transport dingoes to the mainland to dump them. But I was a novice here. Much later it was explained to me: allegedly rangers had taken dingoes targeted for destruction to the mainland in a misguided effort to save them.

Coming to his conclusion Hansen told us that he did hold back some of the gory details about dog attacks. But at the beginning of his talk he had promised to open his heart and tell us how dingoes had affected him on his place. 'It is really hard on your soul,' he said, 'when you hear calves and weaners carrying on in the middle of the night and you go out there and you see calves with holes eaten in their rear ends. These calves are attacked and they haven't even been born, they're only halfway out. You go out there and there's a cow and she's bled to death because she's got no tongue and there's this bloody dog wandering around it.'

Hansen didn't believe in baiting. 'Not only is it a nasty way to die, it has flow-on effects for other animals as well.' But, if farmers 'want to bait and they choose to bait they probably have to bait'. He said the Fraser Coast Council offered a bounty of $40 per dingo scalp and received about 100 scalps a year.

The forum belied my preconceptions of how things would be done in Queensland. In the last session, sitting on the stage together in front of about 60 people, three tame dingoes from a local wildlife sanctuary and Jennifer Parkhurst's diabetic-alert dog Kari, the pastoralists and the dingo advocates did not insult each other. The primary producers were gracious. Perhaps they could treat the dingo conservationists, who sat still like statues, with such courtesy because whether or not they killed dingoes on their property remained up to their discretion. It was, as Hansen said, their prerogative.

Outside the community hall, before the forum, each of the three tame dingoes had given me a quick sniff up and down my jeans to my knees. One, Yerri, even let me pat his big wide head after he had sniffed my arm. Inside they sat and lay patiently through all the words, and, when they were allowed, moved about on their leads, sinuous and silent. The youngest of the three, Spirit, a one-year-old white female, was gentle, quiet and inquisitive, ears pricked, nose forward, leaning in to Kari, Jennifer Parkhurst's little dog, who barked at her. Spirit was completely unperturbed by Kari's fear and bluff.

Spirit did not want to sit still for a photograph at the end of the day. I couldn't blame her. She would have been tired from the artificial lights and smells, bored from being inside for three hours.

Darkness filled the car park while I chatted with Karin and Malcolm Kilpatrick from SFID after the talks. Karin told me that QPWS ranger Linda Behrendorff was at the forum. She had been working for a long time on K'gari and had dingo expertise. She was one of the rangers I had requested to interview. I was still waiting to hear back from Parks about interviewing rangers. Patience. I needed to be optimistic.

It took me a long time to get back to my accommodation that night because, between Pialba and Torquay on the nondescript multi-lane roads, I got completely lost.

* * *

When I called my mother she told me how she and my father had driven to Urangan in her Ford Prefect in 1954, before they were married, with her Aunty Ivy as an ineffectual chaperone. My father went drinking at the Urangan Hotel, where they stayed. Later he drove some of his drinking friends back to their ship, docked at the end of the long pier. There was no room to turn around at the end of the pier – he had to reverse all the way back in Mum's car, over a kilometre.

Urangan Pier runs from the mainland to a deep-water channel in the Sandy Strait and was built from Fraser Island timber in the early 1900s to facilitate the export of sugar, timber and coal. In my parents' time timber was still exported and petrochemicals were imported. By the time I visited, the pier performed no import or export function. It was a tourist attraction, a place of leisure, a place for people to fish and stroll.

The day after the forum I drove along the Esplanade to Point Vernon. It was a beautiful spot, with a view east across Hervey Bay toward the island. North was open sea. Beauty made me curious, so I continued around to Gatakers Bay and parked in the shade down near the shore.

It is inadequate to say that a place is somewhere – all places are somewhere – but this quiet bay with its dark rocks sweeping away from the shore into the benign-looking water was definitely somewhere.

In her essay 'Playing with fire' Australian author Jacqueline Wright describes the monsoonal vine thickets of the coastal Kimberley dunes around Broome as soft places, or havens. These places, with their quandong, walmadany and bauhinia, have been favoured camping areas for Aboriginal people for centuries. Wright describes how Jeanné Browne has spent many years walking the Goolarabooloo songlines of this country, and how 'It takes a great many years of walking the trail to understand that it's not about getting to know the country but, rather, the country getting to know you'.[26]

---

26   Wright 2015

As a child I took comfort from familiar, soft places; they were my confidants. As I grew up I held a nebulous belief that trees and rocks and places and humans could communicate with one another. But in my twenties I started to quash my belief that I could perceive the universe's sentience. In January 1990 on a camping holiday on the south coast of New South Wales, a soon-to-be ex-boyfriend and I drove down a road with dark, regular-shaped plantation conifers on one side. On the other side of the road native, bleached, motley scrub leaned forward like a bunch of eccentrics heckling the straight conifers. Ghostly eucalypts on cleared hills in the distance looked as if they had died for lack of company.

I was unhappy but I did not know how to act in a way that would make any difference. My brother Bruce was HIV-positive. My brother Roderick was in jail. I was losing my faith in my ability to interpret what animist spirits of place might be communicating. In looking for meaning, I had started to think, was I imposing meaning? Perhaps those surviving trees were not throwing their limbs up to the sky in horror at the grotesque shapes of their uprooted companions. And if they were, could I bear to listen to them?

Nevertheless, when other people write about how land communicates and what it says, I find their assertions plausible, even though different people use the information country gives them in different ways.

Queensland dogger Ned Wilson writes in his memoir, *A Dogger's Life*, about the agency of particular places as he decides where to put his traps to catch dingoes:

> you begin to work instinctively to pick out good places to set traps. Where another man would take no notice of this certain spot and go past it and set a trap somewhere else, it will attract you. This place could be a rock, tree, tussock of grass, or small shrub. It seems to say to put a trap there.[27]

Gatakers Bay felt hospitable to me: calm water, a safe landing place for boats. I felt as if I'd arrived somewhere. I suppose all animals, including

---

27   Wilson 2001, 54

humans, are attuned to habitat, places we might inhabit. But perhaps this sense of familiarity came from some other source – no source that I can give references for, no source that people would find credible as evidence. But as those rocks, trees, tussocks and shrubs spoke to Ned Wilson, this place stopped me, made me notice it.

As my mother tells it, my parents' trip to Urangan was a long car holiday. Like many of their generation and generations after, my parents went to Queensland for a break. But now I wonder whether my father was conducting his own unspoken search. He didn't go for the fishing, even though when I knew him he liked fishing, because, according to my mother, he didn't go fishing when they stayed in Urangan. Perhaps he liked sandy waterways. Or perhaps he was drawn there because of his Grandpa Watson's tales of blackbirding.

Grandpa Watson was my father's mother's foster father. When my father was a boy Ma and Pa Watson lived on Harris Street in Pyrmont, which was then a quarried, industrialised sandstone peninsula surrounded by docks just west of the city of Sydney. Grandpa Watson told my father stories about being a pilot on Sydney Harbour, as well as blackbirding. It is hard for me to check his identity because, when I finally got around to asking my father what Grandpa Watson's first name was, he couldn't remember.

'Watson,' he said and paused. 'Grandpa Watson.'

Even Pa Watson's last name comes with some doubt because, according to Dad's mother, Grandpa Watson stowed away from England to Australia with a friend (the image in my mind is of two boys about the age of fourteen) and, for some unknown reason, they swapped names. His name does not appear on any of the old birth certificates or documents I have.

At Gatakers Bay there is a small plaque commemorating the 12,000 Pacific Islanders brought to the Hervey Bay region between 1863 and 1906 to work as conscripted labourers in the cotton and sugar cane fields. Some of the Kanakas, as they were called, lived at Point Vernon and some of them died there. Fifty-five Islanders are buried in unmarked graves at the Polson Street Cemetery in Point Vernon. The death rate for Islanders on Queensland plantations in the 1880s was about thirty-five per cent, which, when compared with the death rate for Australian prisoners of war in Japanese detention in World War II at around thirty-six per cent,[28] shows how little their lives were valued in the great endeavour of turning land into profit.

Grandpa Watson's stories made an impression on my father because he told us, at family gatherings like Christmas lunch, how when he was at primary school in the 1930s his classmates did not believe his stories about the Pacific labour trade, one of Australia's versions of slavery, until one of his teachers verified that, yes, it had happened. My father had wanted to join the merchant marine but instead, in 1944, after he sat his Leaving Certificate for the second time, he joined the Royal Australian Navy as an able seaman.

Now I wonder what kind of hospitality Gatakers Bay offered to the South Sea Islanders and their overseers. But when I was there it was instinct not ethics that guided my apprehension of this place. Its familiarity, its inhabited-ness, made complete sense to me when I read the plaque and thought about my father.

Or perhaps when Dad drove in reverse down Urangan Pier in 1954 he was rehearsing his own future. More than thirty years later with his second wife Judy he drove around Western Australia to see the wildflowers. He joked he'd driven 5000 of the 10,000, or whatever, kilometres they travelled in reverse because Judy kept saying 'Stop, go back' so she could photograph a particularly beautiful stand of flowers. Later still, when Dad was shrunken and dying in hospital he had waking dreams of driving. In an opiated haze his bony hands grasped the wheel of an imaginary car. In his delusion perhaps he really was still in control.

My obsession with dingoes, I'm coming to realise, is something to do with authority and family, habitat and the unquiet dead.

* * *

On the way to Rainbow Beach, to interview Jennifer Parkhurst, I passed through Maryborough, a grand old dream of a town. One of its founders was a trader and publican called George Furber who built a wool store and a wharf on a stretch of river that was called Moonaboola by local Aboriginal people and was named the Wide Bay River by

---

28   Evans 2008, 195 and n. 10

explorer Andrew Petrie in 1842. The Aboriginal people of Wide Bay resisted white settlement in many ways, including by stealing sheep and attacking settlers.

On 13 October 1847 in a dispute over rations Furber received an axe wound to the back of his head from two Aboriginal men who had been stripping bark for him.[29] The men 'decamped, taking with them two double-barreled [sic] guns ... and nearly all the tools with which they had been working'[30] and Furber rode 240 kilometres for medical help in Ipswich.[31] By December Furber was back and the first shipment of wool from inland stations was loaded onto a schooner at his wharf. A few months later, he shot dead an Aboriginal man with his pistol at Henry Palmer's store in Maryborough.[32] Furber was accused of killing at least three Aboriginal people.[33] But the vendetta was finally avenged in 1855 when Furber and another settler were ambushed while they were sawing timber in the scrub near Tinana Creek. Both men were killed, Furber with his own axe.[34]

In 1847 the river was renamed the Mary River after Governor Charles Fitzroy's wife, Mary Lennox, who, although we share a name, is no relation that I know of. More Europeans arrived to set up wool stores, inns and wharves. By 1851 nearly 300 shepherds, stockmen, loggers, sawyers, axemen, bullock drivers, labourers, traders, manufacturers and their families lived in what was at the time the northernmost British settlement in the colony.[35] The town was moved downstream in the early 1850s and, by the looks of the substantial 19th-century buildings that were once banks, warehouses and bond stores, it enjoyed prosperity.

Visitor information was in a corner of an imposing brick building set back from the street on a grassy rise. Built in 1908, City Hall and the substantial staircase I ascended to its front portico seemed to be testaments to civic pride and optimism. I browsed the tourist

---

29 Evans and Walker 1977, 50
30 *Moreton Bay Courier* 1847, 2
31 Evans and Walker 1977, 50
32 Evans and Walker 1977, 50
33 Gardner 2015, 5
34 Evans and Walker 1977, 50; Lynch and Lunney 1996, iv
35 Lynch and Lunney 1996, v; University of Queensland and FAIMS 2014

information centre looking for a decent map or anything else that was interesting. The woman at the counter asked me what I was looking for, where I was going. Her deep-set steely blue eyes and her intonation reminded me of our Queensland cousins – decorous, intense, intelligent. I was evasive about what I was looking for and where I was going because she gave me the impression that she would tell me what she thought I needed rather than letting me find it for myself.

Down at the Mary River Marina on the day of my visit the Muddy Waters Cafe was closed and no traffic ran on the rail that once carried sugar, timber, wool, tallow and hides between the town and the port. According to Maryborough port's website, in the last half of the 1800s the town was second only to Sydney as a port of immigration.[36] Between 1862 and 1901, 20,000 immigrants from Britain and Europe arrived in Maryborough in addition to the 12,000 Pacific Islanders.[37] But I saw no people. A handful of yachts were berthed parallel to the jetty that ran parallel to the riverbank and a sign advertised 'The Wharf – Now Leasing – Marina Berths, RV Park, Laundromat, Photographer, Cafe, Shop and Office Spaces'.

On the side of a cavernous corrugated-iron shed on March Street, which ran up from the river, another sign asked passers-by to imagine what appeared, by the sketch, to be a development of four blocks of six-storey flats arranged around a big, flat, paved open space. In the foreground of the picture was a wide set of stairs with very low risers. On one side of the stairs was drawn a building that looked like a watch tower, but which must have been an elevator.

There was something incongruous and deeply funny about this sign, asking me to imagine a bland urban development, fixed to the side of a building more marvellous and mysterious than anything I could imagine. The scale of these Queensland buildings, of the Queensland coast, of the amount and size of timber here awed me. Along Kent Street, which becomes the road to Cooloola, massive timber sheds, once shipyards, groaned in dignified ruin. Unreal. They had to be real. My imagination couldn't have built them.

---

36 Maryborough portside, 2006–18
37 Lynch and Lunney 1996, 1

Over lunch at a cafe in Richmond Street, about a block from the well-maintained Victorian Maryborough Court House, I read an email from Senior Conservation Officer Naomi Stapleton informing me that QPWS were focusing for the time being on projects that were guided by and had been successful under the Fraser Island Dingo Conservation and Risk Management Strategy so they could not participate in my research. Some time later I noticed that, inadvertently in her email to me, Stapleton had passed on Senior Ranger Ben Steep's email to her, in which he commented that he thought that this research would confuse the demarcation between staff members' personal thoughts and motivations, and their role as employees in the Queensland public service. He was absolutely right.

3
Coolooloi

women … are prone, not just to disobedience or theological error, but also to flirting with animals
    Anne Enright, 'The Genesis of Blame' in *London Review of Books*, 2018

Rainbow Beach is on the Cooloola coast, named for *goolooloi* or *coolooloi,* the Gubbi Gubbi/Kabi Kabi Aboriginal word for cypress pine (*Callitris columellaris*).[1] Cooloola is such a euphonious word but the wind was not singing to me through the cypress pines' spindly foliage as I drove south. My first impulse is to describe the road as unfrequented and featureless. But there were features. The ones I remember were controlled by state and federal government agencies and contributed to my sense of disorientation and unease: monocultures of dark green, upright Caribbean pine were growing in uniform rows at the Tuan and Toolara State Forests; the flora in the Wide Bay Department of Defence training area was less regimented but the land was still fenced off, closed to the public.

    I thought that I could be dispassionate about the way the farmers at the forum the day before had described dingoes' predation on cattle and calves, and the impunity with which they had spoken of their own

---

1   Bell and Seed 1994, 43

violence toward dingoes. I thought I could take this violence in my stride; keep my composure. But researching dingoes had taken me into a dark world of fear and killing, the subconscious of the Australian settler-colonial psyche. I expected my interview with Parkhurst would traverse similar territory. My obsession was nothing compared to hers but I thought if she could tell me why she was obsessed I might know more about why I was.

When I arrived at Debbie's Place, my accommodation in Rainbow Beach, the reception desk was staffed by Suzie, a flighty, friendly little long-haired dog wearing a nappy. 'She's special,' Debbie, a human, explained as she took over.

Jennifer Parkhurst's house wasn't far from Debbie's Place but I drove there because I didn't know what time I would be leaving and, mindful of snakes and men, I didn't want to walk through the unkempt scrubby patches of Rainbow Beach on my own in the dark.

Parkhurst was tall and she looked strong but I knew she had many health problems, including type 1 diabetes and fibromyalgia, a condition whose symptoms include widespread pain and tiredness. Now I'd made the 120-kilometre journey from Hervey Bay to Rainbow Beach, I understood better how tired she would have been after travelling to and from the *Dingo – friend or foe?* forum. I commented on how well she looked and she explained that sometimes that was part of the problem – she looked well so people assumed she was well. We sat on the patio out the back of her compact, two-storey A-frame house. Late afternoon darkened to evening as we talked. Sometimes Kari, her diabetic-alert dog, sat on her lap. Kari was a soft-haired poodle–Maltese cross: intelligent, a fussy eater, and, on the surface, not at all like the dingoes who had captivated the person she cared for.

Parkhurst was fascinated by dingoes' supple grace long before she came to the Cooloola coast. When she found out there were dingoes on nearby K'gari, she wanted to learn as much about them as she could. Whatever she found out she wanted to share with people; she wanted to teach the world about dingoes. She had to win the dingoes' trust so they would incorporate her into their pack structures, which she managed to do, she said, 'by being submissive'.[2] She ignored all she'd been told about

---

2   Parkhurst 2015b

how to approach dingoes and worked on her understanding that canids have alpha and beta and gamma animals in a pack; that they rely on a system of dominance–submission.

She was in their territory, she told me, she didn't want to impose on them. 'I would squat down and let the dingo suss me out. Spend a short time with each different pack that I was observing until slowly, over time, I'd stay there a bit longer and a bit longer and a bit longer until it got to the stage where they expected to see me there and, eventually, they would greet me like a member of the pack when I did turn up.'

For six years, Parkhurst did her research 'without using the stuff the scientists were using, the ground spray things, the tracking collars, the GPS stuff, the cameras in the bush and all of that. I wanted to be able to prove that I could do it without luring dingoes to me with food.'

She believed that dingoes have emotions, which, she said, people refuse to acknowledge so that it is easier for them to hurt, trap or kill them. She told me how their social rituals demonstrate that dingoes love one another. She witnessed family bonds strengthened by greeting ceremonies in which every member of a pack greeted every other member, even if they had been separated for only half an hour, each dingo wagging their tail and licking the muzzle of every other dingo individually.

Once she found herself in a 'gang war', when a pack she was observing decided to go hunting in another pack's territory. The pack she was following didn't know the home pack were waiting in ambush. When the home-range dingoes decided enough was enough, that they could tolerate no further encroachment, they 'all jumped up and started, buffo! And there were dingoes running either side of me, all around me, all the rest of it, with their fangs bared and all that kind of stuff. And I was just, "Oh, I've got to get photos of this."' Her deep laugh conveyed the adrenaline rush of teeth, paws, fur and dingo-focused mayhem. 'They totally ignored me,' she said. 'They were involved in their thing.' In almost all cases, she said, territorial fighting on K'gari is posturing and not injurious.

She told me how honoured she was to be permitted to see wild dingoes mating in the bush. 'Dingoes not having emotions! The foreplay went for an hour. It was just gorgeous to watch. It was all so human. The way she teased him up and all the dancing, jumping on top of each other and prancing around and playing and all the rest of it. And then the copulatory tie.

'Afterwards, they loved each other. The emotion and the affection. They were licking each other's muzzles and faces. They were rubbing bodies, spiralling around each other and under each other and rolling around. It was such a special moment. And I thought, "This goes on in the wild all the time and nobody sees it."

'You assume animals mate because that's instinct and they have to do it. But actually there's this whole emotional side to how they go about mating.'

Dingoes, she told me, do share territory and resources but there is a ritual around how intruders must behave when they enter other dingoes' territories. Whether it's a pack or whether it's an individual, the dingo or dingoes seeking access to another group's home territory approach the home-range dingoes and stand within so many metres of them with their tail and ears down, waiting for a signal that tells them they can advance closer. This process might be gradual, until the newcomer is close to the home-range dingoes, who circle and sniff.

The home dingoes might grab the intruder's neck, throw them on the ground. They might nibble the newcomer's tummy, their most vulnerable part. The intruder postures to signal submission and, by baring their tummy, Parkhurst explained, they are showing trust, demonstrating that they are not there to fight. She admitted that sometimes these rituals get rough but, generally speaking, she said, once the intruder displays submission the home-range dingoes allowed them to enter and pass through.

Parkhurst told me that Aboriginal people have similar territorial entry ceremonies. An intruder needs permission, she said, needs to wait until the elders acknowledge them and invite them to step closer. 'You don't just wander in to somebody else's territory.'

I was fascinated by the form and ritual, the etiquette of dingo social life. In open country, Parkhurst told me, dingoes travel in formation in order of rank, with the alpha male and female leading, and subordinate animals behind them. In the bush, she said, formation wasn't so important; everybody scattered to get through the undergrowth. She described how packs are usually made up of the two alpha individuals who breed and whichever dingoes survive and stay around from the previous year's mating. These dingoes become the aunties and uncles, or alloparents, who help to raise the current year's pups.

Parkhurst was not a disinterested observer. She named the dingoes she was observing. She knew them as individuals. She knew their personalities, identified them by the way they walked. Some of them became close friends. When dingo pups died of starvation in the lean time after they were weaned and before they had learnt to hunt for themselves, she grieved.

People told her, 'Get used to it. Detach yourself.' But, Parkhurst said to me, 'I could never detach. And I think that's kind of good, even though it hurt. It gave their lives meaning, to me.'

As well as dying of starvation, she said, pups were being 'knocked off'. She believed that QPWS found the island easier to manage with fewer dingoes. The mortality rate for young dingoes at that time was as high as ninety per cent, she claimed, because the playful, boisterous, rough behaviour of young dingoes was misinterpreted by the public. Their play involves body slamming, jaw sparring and chasings or tip. The time young dingoes start to become more mobile and to come out onto the beach coincides with peak tourist seasons on K'gari.

'All of a sudden there's people everywhere and they get excited and they don't know they're not supposed to approach people,' Parkhurst said. 'They don't know people have been told they're dangerous. So they'd go bowling up to somebody and they'd grab them and give them a nip on the leg or they'd chase around or they'd do a play bow, which people are told is dominance testing but actually it's just an invitation to play.

'They are hungry and of course they're scavengers so if they came across people in a campsite, and the campsites are placed quite often right at the entrance to their territories, you've got these little pups trying to get back to their territory and there's suddenly people there and they don't know who you are or what you're doing.' For a time dingoes were not allowed on the beach between 5 a.m. and 5 p.m. 'How,' she asked, 'do you tell a dingo, "You're not allowed on the beach"?'

Parkhurst became openly critical of QPWS's practices of hazing and tagging dingoes, their system of incident reporting, and the review processes of the dingo management strategy. She explained how QPWS encouraged people – including tourists and residents of the island – to report any dingo sightings to them. These sightings were recorded by rangers on a dingo incident or interaction report. The incident or interaction was coded from A to E: Code C was a dingo who has been seen around people too often; Code D was a dingo who approached people; and Code E a dingo who nipped people. The system generated a lot of literature. At the Save Fraser Island Dingoes office in Hervey Bay the day before I had seen a stack of paper nearly half a metre high. They were incident reports on dingoes for the last six months, which SFID obtained through requests under Queensland's *Right to Information Act (2009)*.[3]

Parkhurst claimed that the reporting was too subjective, that each ranger perceived dingo behaviour differently and dingo behaviour was increasingly misunderstood. 'When incidents occur eyewitnesses aren't given the opportunity to report,' she said, 'because we've never got one through our Freedom of Information searches. Never ever got an eyewitness statement. It's always a ranger's statement about what happened. And it's always the dingo's fault.

'On many occasions we're able to track down the people or the witnesses and find out what really happened and it's not what's in the official incident reports. It's scary. Like the dingo that was killed recently, there was a child floating down Eli Creek and the dingo reached out apparently and grabbed their hair. So what? It wasn't an attack. Dingoes love water play. They're excited by water. If the people are in water and splashing about they want to get in there and splash about as well.

'An attack, by their own literature,' she said, 'is a dingo biting numerous times and unable to be deterred. Well, that's never happened. Dingoes come up and nip and bite. They puncture the skin but they don't rip it off. So QPWS's language needs to be explored and they need to better understand how they're approaching the public.'

In Parkhurst's view, incident reporting led to mistreatment of dingoes. 'After a while if a dingo's got enough reports made against it, it's

---

3   Queensland's *Right to Information (RTI) Act (2009)*, which replaced the Commonwealth *Freedom of Information Act (1992)*, requires the Queensland government's Department of Environment and Science (DES) to publish a disclosure log on its website, which provides access to information that has been released in response to completed RTI access requests. See Queensland government (2017–2018).

classed as a habituated animal and it's put on watch. Habituated animals are quite often hazed more regularly than other animals. Hazing does not have the desired effect of keeping them away from people. Hazing can actually change their behaviour and make them – not aggressive, I would never ever say I've met an aggressive dingo – but it can make them more defensive maybe, when they see people.'

Hazing is a practice 'designed to deter dangerous wildlife from frequenting certain areas and re-instil in them a natural fear of humans'.[4] When rangers haze dingoes 'for their own protection', as one Queensland government ecologist put it to me, they attack them with clay pellets from a slingshot or fire 'ratshot' projectiles at them from shotguns.[5] Parkhurst was sceptical about whether hazing had ceased, as the most recent Fraser Island Dingo Conservation and Risk Management Strategy (FIDCRMS) claims,[6] because, she said, rangers had a habit of doing it out of uniform in unmarked cars out of hours. Even though QPWS's public statements say the practice has been stopped, the most recent dingo management strategy does not prohibit hazing completely:

> Physical hazing of dingoes is no longer practised as a management intervention unless under an authorised program with the primary purpose of evaluating the appropriateness in minimising the need for euthanasia of animals that otherwise may become habituated.[7]

According to Parkhurst, juvenile dingoes became classed as habituated animals and accrued 'black marks' any time they were seen with people. 'All they've got to do,' she said, 'is make some kind of contact with a person and they're killed.'

Ear tagging, in her view, was another practice that contributed to dingo mortality. She observed that as soon as the pups were ear tagged they lost their ability to hunt because the tagged ear drooped. When the

4   Ecosure 2012, 89
5   Ecosure 2012, 89
6   Ecosure 2013, 5
7   Ecosure 2013, 11

pups are about three or four months old their parents try to teach them to hunt but, Parkhurst said, they need both ears to be able to locate a sound.

Ear tags delaminated through weathering, and wear and tear, so that the colours came off, making it impossible for even experts to identify dingoes, let alone a person who was in a state of panic because a dingo had approached them. The rangers didn't need the ear tags to identify dingoes, she claimed, because it was easy to identify a dingo if you knew what you were looking for. The tags, she said, were for the public, so that they could try to identify dingoes for the incident reports.

Parkhurst claimed tagging is not only ineffective and unnecessary but cruel. She described how dingoes ripped out their ear tags and how the tags got caught on vines and other foliage in the bush. Dingoes, she said, ended up with horribly scarred ears. She thought QPWS would never stop ear tagging, though under the latest management strategy young pups were spared. The 2013 management strategy recommends only dingoes with a 'minimum weight of approximately 10kg, of reasonable body condition [...] that exhibit, or are anticipated to exhibit, problematic behaviour' be targeted for ear tagging.[8]

Since late 2002 Parkhurst had seen the management strategy 'stepped up' but she thought the review process was inadequate. 'They just keep reinforcing their own stuff.' She was disappointed with the most recent review, published by an organisation called Ecosure in 2012: 'It was supposed to be scientifically independent. The people that did the review were the people that were involved in the implementation of the first-ever strategy. It wasn't independent.

'They think their management is fine,' she said. 'They indoctrinate new rangers into the management strategy and the new rangers believe them. So you've got this culture where all of the people in management on the island either believe it or have to act on it that dingoes are inherently aggressive and that you have to keep dingoes away from people at all costs. That dingoes should be – it's ironic, they say, "returned to a wild state" – but then they say they're inherently wild so therefore we shouldn't interact with them.'

It is clearly very hard for some animals to conform to human definitions of 'wildness'. If a 'wild' community of animals is 'separate

---

8   Ecosure 2013, 5

or independent of human contact' in respect of geographical location and human influence,[9] how many animals have the freedom of independence, in location and influence, from humans? The dingoes of K'gari occupy 184,000 hectares. Their territory is visited by 400,000 tourists a year. Some of them have become what political philosophers Sue Donaldson and Will Kymlicka describe as 'liminal animal denizens',[10] meaning that they are in between, they do not fit into the domesticated/wild dichotomy. Such dingoes live independent lives, hunting wild prey and eating what they find, but they are also accustomed to people and curious about them and their accoutrements. On K'gari some of them seem to be trying to adapt humans to their needs. The term 'liminal animal denizen' is, to my mind, an unsatisfactory and anthropocentric label. 'Liminal' connotes living on the edge and 'denizen', in addition to meaning resident, describes a naturalised plant or animal, established in a place where it is not native. Further, and even more demeaning for dingoes on K'gari, a denizen is an alien, or an individual permanently resident in a foreign country, who is granted specified (and limited) citizenship rights. I doubt dingoes would see or describe themselves this way. Liminal animals are not well integrated into human ideas of how animals should act, and how harmonious relationships between humans and animals can be achieved. But as human populations increase, and encroach on and destroy animals' habitats, growing numbers of wild and not so wild animals could be described as liminal animals. In this field dingoes are adaptive leaders in how they have managed to co-exist with humans in places like K'gari. They might think of themselves in terms more like how jazz maestro Duke Ellington conceived the highest form of genius – as 'beyond category'.[11]

Strictly enforced separation between humans and dingoes does not recognise the semi-domestic relationships Aboriginal people have had with dingoes for thousands of years, Parkhurst observed. She preferred the way the Department of Forestry managed dingoes: people fed and

---

9   Hadley 2015, 92
10   Donaldson and Kymlicka 2011, 210–51
11   Tucker 1995

wormed them, and there were signs saying, 'Leave your food scraps for the dingoes.'

She started writing articles and voicing her opinion about ear tagging, hazing and the killing of dingoes. She found out that the QPWS dingo management strategy had been put in place seven years before Clinton Gage was killed in 2001. She blamed QPWS management for causing immense food stress for the island's dingoes and she pointed the finger at Parks: 'I was starting to say, "You caused the death of Clinton Gage. Your management."'

* * *

Parkhurst is well known because she was prosecuted for feeding dingoes on K'gari, an action she says she took to stop pups dying of starvation. 'I'd sat with them while they died. It was almost as bad for me watching them die as it was for them dying. It was this huge emotional thing. Why don't I just feed the bloody things? Why don't I just feed them and then they won't be suffering? Because starvation is a very slow and painful death. It's a terrible death. But I was determined I wasn't going to do it. I wanted to be able to prove that I … you know.' She trailed off.

On the internet I found film footage Parkhurst had taken of a pup she named Pepper. In the footage Pepper lies on her side with her neck extended and her head turned around at a strange angle. Her eyes are delirious looking, neither closed nor open. The side of her torso rises and falls with breath that is coming and going through teeth that look clenched.[12] What could be more different from Pepper's *agonia* than my dog's anxious, belligerent exuberance around the meal time that she knows will come every night?

My dog's mother was a well-behaved cream kelpie brought home from a pet shop. Her father was a wandering cattle dog who came to stay and reached sexual maturity before his human hosts were expecting it. Both of these Australian working breeds have dingo ancestry[13] – acknowledged in the case of the cattle dog and not officially

12   Parkhurst 2012
13   Parr et al. 2016, 184

acknowledged in the case of the kelpie.[14] My dog Zefa is a thoroughly domesticated dog – cooperative and highly motivated by food. Her interest in food starts when she sees me in the kitchen in the morning but becomes more intense while I'm preparing dinner in the evening. She lies close by, right in the middle of the entrance to the kitchen, and watches me attentively. It looks like love. She starts pestering for her dinner while we, her human family, are eating ours. First she targets people who in the past have dropped scraps for her. She rests her head in their laps; she crowds them, lying close to their chair; she noses their elbow. Her ears are forward and her eyes are alert. She looks like a puppy again. Eventually she might lie down under the table, or under my chair so that I feel like I've laid a dog egg.

If we touch her while she's lying down she springs up and addresses us. She won't let us touch her. In the last few years she has started to bark at us when we are almost finished eating. If we ignore her, turn our heads away and cease to make eye contact with her she quietens down, lies down again. But as soon as we turn to her, or touch her, she is up again, insisting loudly and impatiently that it is her dinnertime too. When someone carries the dishes from the table to the kitchen she follows them, prancing with her tail high and her nose up to savour the smell of food scraps. I love the parade of dog following dishes to the kitchen after dinner. I imagine her as a wild canid, frisky with hunger, even though she's never been as hungry as a dingo because dingoes are, as pastoralist Lindsay Titmarsh put it, 'used to being hungry'.[15] I can see her excitement, hear her calling up her packmates to hunt. If we go outside at this time of day she runs around me and barks at me. I plant my legs wide, bend my knees, open my arms in a Steve Irwin stance. I say, 'Isn't she a beauty? Let's catch her!' She stands in front of me, shoulders lower than her rump, front legs splayed in a play bow. With my feet planted I move my upper body from side to side in front of her. She turns her head to one side, then suddenly bounds away. She circles me, veers into me, runs circles and changes direction quickly. I haven't got a hope of catching her. This is the energy, I think, that would help

---

14   See Lennox 2013
15   SFID 2015 (Lindsay Titmarsh quoted)

her catch her dinner if she had to. But my dog is not a dingo; a dingo is not a dog.

In contrast, in Parkhurst's footage, Pepper lies still. Her siblings rest around her. The next scene shows Parkhurst squatting beside Pepper and patting her. Pepper is sitting with her back to the camera, turning her head and seeming to enjoy Parkhurst rubbing her gently behind her ear. Parkhurst's voiceover says she has never patted a live K'gari dingo before; she is allowing herself to touch Pepper because she knows she is going to die. Parkhurst is crying. In the last footage in this sequence[16] the camera is in Parkhurst's hands as she approaches a pup's body. The pup lies with her neck extended on what looks like a bed of dried casuarina fronds under low-hanging, brittle, grey, twiggy branches. Parkhurst's breath catches. She sobs. 'The poor little thing,' she says, 'It's so unfair.' The footage shows Parkhurst's left hand encircling Pepper's abdomen, flies moving from Pepper's fur to Parkhurst's skin as she says, 'Little girl, what a tough little life she had, eh ... Looks like this might have been a favourite spot of hers ... I thought she'd come to a favourite spot.' Parkhurst's hand strokes Pepper's body. Her voice says, 'Thank you, baby girl, for the hours and hours of joy that you gave me.' Her fingers touch Pepper's paws. 'It's been such a pleasure knowing you.' She fondles Pepper's ears. 'You were a beautiful dog. A beautiful, beautiful dog. You really captured my heart, little girl, with your courage and your love.' Flies buzz. 'You're a really beautiful girl,' Parkhurst says, 'and I just want to thank you. Thank you very much.' Her hand encases Pepper's flank. 'It's been an honour and a privilege to know you.'

When canids hear humans talking to them, form is content. The tone of our voices is the message. Whatever my language means to my dog, she tolerates my saying the same things over and over again. I continue these repetitions almost instinctively as if they could be some kind of antidote to suffering. I wonder sometimes whether Zefa loves me only because I feed her. She expresses her love so genuinely, I don't mind, I'll take that. I'm sure it's not all about food even though food changes everything for a canid. But free-ranging dingoes? Is the most

---

16   Parkhurst 2012, 23.05 mins

base explanation the only one, their interest in humans is simply their interest in food?

According to Parkhurst, DERM suspected she was feeding dingoes all along. They wanted an excuse to remove her from the island, she claimed, because while she was there in the bush she knew where the packs lived and how many pups were being whelped. Her presence, she believed, stopped rangers from helping the mortality rate along by killing pups and destroying dingoes.

Parkhurst believed that QPWS not only wanted to stop her from visiting the island; they wanted her in jail. Perhaps this feeling was unfounded paranoia; perhaps she had reasons to feel paranoid. Her difficult relationship with her then-boyfriend would not have helped her insecurities. She told me he tried to convince her to start feeding dingoes. She said he told her that the dingoes were her family and that she couldn't let them starve. She had seen so many litters of pups starve, he said, why didn't she feed this one litter and let them grow up, and she would have the satisfaction of knowing that just one litter out of all the litters she had observed had survived.

Well, hell, why not? Parkhurst thought. *The QPWS hates me anyway. They think I'm feeding them. Why don't I?'*

He started the feeding, Parkhurst said, and she told him to feed only the alpha female away from the pups so she in turn could provision the pups and so the pups would not be seen as habituated. This alpha female was a dingo who had befriended Parkhurst and 'led' her to her pack in the bush.[17] Parkhurst called her Kirra, a name she said was derived from two Aboriginal words, one meaning 'sent from the sky' and one meaning 'blue water lily', because Kirra loved water. The feeding was only in the last six months, she said, and it only started when the pups were four months old.

Kirra and her pups belonged to a group that Parkhurst called the Hook Point pack. Their territory stretched across the southern tip of the island from the eastern beach and Jabiru Swamp behind the beach to the tangled forests of Coolooloi Creek, facing the mainland. Kirra had visited camping areas in search of food since she was young without incident, Parkhurst told me, and the pups, too, realised they could get

---

17   Robson 2013

food from humans. Parkhurst believed it was her relationship with Kirra that brought this dingo and her family to the attention of QPWS staff.

'We were seeing her walking out of the bush with a pack of five or six dingoes at her heel,' ranger Colin Lawton said on an *Australian Story* program about Parkhurst that aired in 2011. 'You know that's not normal.' According to Lawton, the dingoes' interest in Parkhurst was 'a food condition response'. 'We started to have reports of those dingoes interacting with people, and aggressively interacting with people … we're getting some really serious attacks on people down there,' he said.[18]

According to a database called the *Humane destruction database for dingoes on Fraser Island*, on 7 May 2009 rangers killed a ten-month-old male dingo at Coolooloi Creek. The next day they killed another ten-month-old male and two ten-month-old females at the same place.[19] These were dingoes Parkhurst had been involved with. Parkhurst thought that the young Hook Point dingoes were destroyed not because they were aggressive toward people but because DERM didn't like her.

She and her boyfriend were not getting along. She believed that his persuading her to feed dingoes was premeditated: 'I think he just had this plan in the back of his mind that, if I can get Jennifer to do this then I'll be able to control her.' I do not know whether he hatched a plan and executed it or whether he took opportunistic actions as circumstances allowed. I think it is plausible that he was jealous of Parkhurst's relationship with Kirra, whom she described as 'like my best friend',[20] and the other dingoes, and possibly tried to find a way to insert himself into their social relationships. Perhaps Parkhurst's belief that her boyfriend wanted to trap her was paranoia. When Parkhurst tried to break up with him, she told me that he made threats and she was scared. She made reports to the local police about him. Not long after he finally realised that the relationship was over, she said, her house was raided by five DERM officers.

They came at 7 a.m. on a warm morning in late August 2009. Accompanying them was a local policeman who knew Parkhurst from

18   *Australian Story* 2011 (Colin Lawton quoted)
19   Allen et al. 2015, supplementary material
20   Robson 2013 (Jennifer Parkhurst quoted)

the reports she had made about her ex-boyfriend. 'The local policeman was very uncomfortable,' she told me. The DERM officers spent the next six hot hours combing through her possessions. Three hours upstairs and three hours downstairs.

'I wasn't allowed to move. I wasn't allowed to go to the toilet without supervision. I wasn't asked if I wanted breakfast or anything. You know, I'm diabetic; my blood sugar was going up. I hadn't brushed my teeth, I hadn't brushed my hair. I was really humiliated and personally embarrassed. To have somebody standing outside the toilet.'

They went through every notepad. They tipped the lounges over and photographed Parkhurst's numerous paintings of dingoes on the walls as well as her signature on the paintings. They took her photographs of dingoes off the walls and they took her book contract[21] off the fridge. They took the hard drives of her computers, which contained her research, her photographs of K'gari dingoes from the last seven years – as Parkhurst put it, 'all I've ever lived for' – and all her family photographs and all her private emails and letters. They took her journals. They took her catalogues of the hard drives, too. 'If they take the catalogues of my hard drives, how do I know I got back everything that was originally on the hard drives?' she said.

The possible ramifications of the confiscations were serious not only for Parkhurst. In Queensland it is illegal to keep dingoes. 'A lot of people would contact me and say, "We've got a dingo we need moved, we've found a dingo, can you hook it up with someone?"' After the raid, Parkhurst's friends, colleagues and contacts were worried DERM was going to raid them too. Now, six years later, she was telling me she still could not spend a long time in the study upstairs because of the trauma of that day.

During the raid officials told her that there were offences they were investigating. She was charged under Queensland's *Nature Conservation Act (1992)* with interfering with a natural resource. 'I was sitting at dens,' she told me. 'I was looking after the pups. I was incorporated into the packs therefore I was interfering with the dingoes which were a natural resource. However, I had a permit to do that. I legally was allowed to do what I was doing.' Despite its prominence

---

21   Parkhurst 2010

in the publicity about her, the feeding, she told me, was not the main charge; she said it was a 'by-the-by kind of thing' because the penalties for feeding were not as severe as the penalties for interfering with a natural resource.

In court in Maryborough on 3 November 2010 Parkhurst pleaded guilty to all forty-six charges related to her interactions with dingoes. These included feeding dingoes on thirty-four separate occasions over a six-month period between 4 July 2008 and 31 March 2009, which were offences under section 118(1) *Recreation Areas Management 2006 (RAM) Act* (Qld), and eight charges of disturbing dingoes in a recreation area from 26 October 2008 and 31 January 2009, which were offences under section 119(1) of the RAM Act. For these offences she received a $40,000 fine. For four charges of interfering with dingoes in a protected area between 29 July 2008 and 12 August 2008 and 18 October 2008 and 22 April 2009, which are offences under section 62(1) *Nature Conservation Act 1992* (Qld), she received a jail sentence of nine months wholly suspended with an operational period of three years. Parkhurst's ex-boyfriend, who gave evidence against her in court, received a fine of $2500 dollars with no conviction recorded. Magistrate John Smith, who had been state coroner in 2001 when Clinton Gage was killed, warned Parkhurst to keep away from Fraser Island's dingoes.

After the trial she went to Victoria, where she had grown up, to stay for three months. Coming back to Rainbow Beach was hard. Looking at the island made her cry. 'To be sitting here,' she said, 'knowing Kirra was over there and I couldn't explain to her why I wasn't with her anymore was heartbreaking.'

* * *

Just before her trial Parkhurst visited the island with her father to say goodbye to Kirra. She didn't know what the outcome would be, whether or not she would be going to jail. They didn't find Kirra so they cleaned up rubbish on the beach instead. As they were driving along the beach to catch the barge to return to the mainland, Parkhurst saw a dingo in the distance. When she told her father it was Kirra he asked her how she knew. She said, 'I know her. That's her.'

They were in a different car, with a different smell. Parkhurst told her father to close his window because she didn't want Kirra to smell her or come near, or to start hoping that she had come back. 'But,' Parkhurst told me, 'she knew. She came straight over to the driver's window and just looked up at me and I wound down the window and I wanted to get down and greet her properly and rub her nose and tell her I loved her. But I just had to be content with knowing that I was seeing her.'

Kirra stared at Parkhurst.

Her father, with tears pouring down his face, said, 'She knows you.'

The barge was approaching. Her father told her that they needed to go, she needed to say goodbye. But Parkhurst could not drive away. She could not say goodbye. All she could do was continue to tell Kirra she loved her.

'Kirra was the first to walk away,' Parkhurst told me. 'She got up, walked off down the beach and sat down. That was her way of saying to me, "You've got to go."'

So Parkhurst drove off. When she turned around Kirra was gone.

We were coming to the end of our interview. Kari, named for both Kirra and K'gari, was dozing. The dark around us vibrated with the sound of crickets. Grief for many losses lives in the story of Parkhurt and Kirra's last encounter – including Kirra's loss of her 2008 pups and Parkhurst's loss of her mother, who died relatively young at the age of fifty-two. To me the sadness felt bigger than Parkhurst or me, and the dingoes on K'gari. I was sad about humans' loss of connection to and lack of willingness to understand, on their terms, animals like dingoes who are curious and sociable.

Why were Parkhurst and I so interested in dingoes? What did we seek to learn? Dingoes' sense of their selves seems different from our human sense of our selves; they are connected to their environment and one another in ways we humans are not. I admired their ability to cooperate with one another, and qualities that I perceived as generosity and a capacity to forgive.

'They forgave me for a lot of stuff,' Parkhurst said. 'I'm really grateful to them for that. And they show forgiveness, too. Kirra would have to chastise the pups. She has to, that's her job, she's the mum, but after she would always give them a kiss, saying, "I'm sorry, sweetie." So they do have this big capacity for forgiveness and love.'

For a long time I underrated forgiveness. Perhaps it is a coincidence but it was only after I had children that I realised that forgiveness is a gift both to the one who is forgiven and to the one who forgives. Nevertheless, I don't know whether, in what circumstances, I am capable of forgiving. Parkhurst described Kirra's forgiveness as a maternal quality. Was Kirra asking for forgiveness as well as forgiving her pups? It's a big generalisation, but are maternal bonds capacious enough to allow effortless forgiveness? Perhaps. Maybe. I don't know. I think of my mother, who was much more forgiving than my father. I did not realise I would fall in love with my children. I did not expect to feel such euphoria about their being in the world, to enjoy their company so much, to gain such energy from the fact that they exist. Not that it is all euphoria, of course, but the euphoria is one of the miracles, one of the things that is good to think about. The freedom of forgiving might be like this joy. A good thing. Like the pleasure I feel when I am with my dog, whom I trust because I believe she is always honest. If she acts happy, she is happy.

The way Parkhurst was submissive, allowing dingoes to set the terms of her relationships with them, reminds me of another reviled animal upsetting human-centred ideas of agency. In Antoine de Saint-Exupéry's novella *The Little Prince* a fox asks the prince to tame him. If the prince tames him, the fox explains, they will create ties, they will need each other. The wheat fields that they both can see mean nothing to him, the fox says, because he does not eat wheat. But, he observes, the boy has golden hair, the colour of wheat. If the boy tames him, the fox says, 'The wheat, which is golden, will remind me of you. And I'll love the sound of the wind in the wheat ... '[22]

The fox explains how first the prince needs to sit some distance away and the fox will watch him out of the corner of his eye. Day by day the boy will be able to sit a little closer. So the prince tames the fox but when the boy has to leave, the fox tells him he will cry. The prince points out that it was the fox who asked to be tamed and says, 'Then you get nothing out of it?'

'I get something,' the fox replied, 'because of the color of the wheat.'[23]

---

[22] Saint-Exupéry 2005, 59–60
[23] Saint-Exupéry 2005, 61

The fox takes such initiative in his relationship with the boy, I wonder who tamed whom. And who was taming whom when Parkhurst spent time with Kirra and her pack? The famous sentences 'One sees clearly only with the heart. Anything essential is invisible to the eyes' are the fox's secret, which he tells to the prince.[24]

I asked my children what it means to see with the heart. How is it different from seeing with the eyes? They didn't hesitate. One told me that to see with the heart is to truly believe in something. The other told me that it means you take something in and you really like it and you can relate to it. Seeing with the heart is different from seeing with the eyes, one said, because with the eyes you might not take it in and you might not like it. The other said, 'You believe it. You feel it.'

Parkhurst needed to believe in Kirra's forgiveness and love. I, too, need this love that changes everything, this energy and space: light to refract anger; air to move grief. It is a form of seeing with the heart. Coolooloi. It sings with the wind through the cypress pines. It fills the colour of the wheat. It lives in the forgiveness of animals.

---

24   Saint-Exupéry, 63

4
Eurong

Who are you?
    The Who, 'Who Are You?', 1978

The next morning at the Rainbow Beach 4WD Adventure Centre, Jack, a cattle dog, barked at Parkhurst and me defiantly and, according to his sense of territory and propriety, perfectly reasonably. We had come to rent a Toyota LandCruiser. Perhaps I was overestimating the notoriety of Parkhurst's conflict with QPWS when I thought that everybody in Rainbow Beach must have known about it. Nobody at the 4WD place mentioned it. The people there treated us courteously. Were they on her side or just being polite, I wondered. Not of course that anyone should have to take sides. But people do. We filled out our paperwork and I chuckled at a sign that read, 'You bought a Jeep. WTF?'

Parkhurst drove the massive LandCruiser and I went into the QPWS office on the outskirts of Rainbow Beach to organise our national parks permit. Conscious of how divisive her dispute with QPWS had been, while I filled in details of my name and our car registration I wondered whether staff knew who was in the car outside. This was only the second or third time she had been to the island since the court case. She had been warned that if a dingo came near her she would go to jail and so visiting K'gari triggered her post-traumatic stress disorder.

We drove north out of Rainbow Beach on a road that ran through bush along a narrow isthmus of land. Parkhurst stopped so I could photograph signs along the way: 'On Fraser, be a dingo saviour.' 'Their survival depends on your behaviour.' 'Be dingo-safe!' Beside this imperative was a triangular-shaped logo, like an upside-down give-way sign. Inside the triangle was the silhouetted profile of a prick-eared dingo with its jaws open and its upper and lower carnassial teeth visible. Two white stripes scratched out of the dingo's snout were its whiskers. This representation – with its blank cut-out eyes, which make the dingo look a bit like Batman, and its close attention to dental detail – appears on most of the Queensland government signs designed to educate the public about dingoes that I saw on K'gari. 'A dingo is not a pet' another sign said.

We emerged from the bush to see the barge in the distance, a small rectangle in a vastness of sand and water and sky. The thin dark line of K'gari was behind it. The barge looked stuck on, perched on the horizon with its yellow details – the roof of the wheelhouse and the sides of the vehicle ramp – bright against the pale grey watercolour cloudscape. It grew as we drove in the soft sand tracks of other vehicles toward it.

The barge to Hook Point leaves from a sandbar that juts out to the west at the very tip of Inskip Point. The barge business is cut-throat. Parkhurst told me how one barge operator would drive another out of business by offering cheaper fares or by directing all the traffic to their barge. Following directions, we drove up the ramp, parked close to the vehicle in front and climbed down from the LandCruiser. Parkhurst and the deckhand, whom she knew, chatted. I inhaled that familiar maritime smell, sea air and diesel and salt-damp on metal made smooth with many layers of paint. I felt the submarine churning below the deck under my feet as the barge pulled away from the beach. I couldn't tell whether the turbulence in the water was made by the barge's engines or because of the currents.

There were more signs on the barge. 'Don't drive dingoes down' one said, with the small type explaining how dingoes can run in front of cars on the beach, where they catch fish in the shallows and 'scrounge in' washed-up debris; on the inland roads, which they use as pathways through their territories; and near bins whose smell attracts them.

Another addressed dingo life cycles and pack dynamics. 'Pet dogs are passive,' it started. 'Dingoes are different.' The message elaborated on the distinction between the dependent domestic pet and the 'wild,' potentially dangerous dingo:

> Feeding dingoes because they appear to be cute, hungry or thin is not kind – dingoes can kill. Dingoes are wild animals. They need to hunt, mate and secure a territory and their place in their pack. They live just like any other wild animal – lions, hyenas, wolves – where only the fittest survive.

Young dingoes, the sign continued, look 'lean and gangly until they build up their muscles by hunting and patrolling their territories'; some dingoes are 'beaten up by their pack members until they are strong enough to fight back'; 'Weaker or sick dingoes may die – "survival of the fittest" rules their lives'. Its take-away message: 'Fed dingoes can become dangerous to people.'

Another sign, headed with the word 'DANGER', provided information about how to keep food away from dingoes. 'Dingoes that get their food from people may become aggressive,' it said. It warned people always to walk in pairs or groups, never to approach a dingo, and never to go bush toileting alone – 'take someone with you to watch for dingoes'. I guess it was evoking an all-seeing dingo, not QPWS staff, when it asked: 'Who is watching your behaviour?'

Another in this series told people to use tightly sealed containers in a vehicle or shoulder bag for fish catch, bait and berley; to never clean fish at campsites, townships or other prohibited sites; and to bury offal at least fifty centimetres deep and just below the high-tide mark. 'Who is watching your catch and bait?' the sign asked. 'You should be.'

I didn't notice anyone else reading the dingo education signs during the ten-minute crossing.

Parkhurst told me there were cameras all around Hook Point where the Inskip barge comes in. The beach around the landing site was eroded, its quiet broken by the brief flurry of engine and tyre activity as the four-wheel drives disembarked across the metal gangplank and churned through the sand toward the inland road. The tide was too high to make it around the point to the eastern beach. Instead of following the other vehicles, Parkhurst turned left. We detoured to Coolooloi on the western side of the island, the territory of the Hook Point pack. She gestured to the northern side of the road – I couldn't see it but not far away through the tangled, light-filled scrub lay the swamp that was Kirra's territory.

Kirra was born in 2003 and she died in 2011. Parkhurst told me that when QPWS rangers found Kirra's body they said that she died because she'd lost a fight with another dingo. She was startled to read a note on Kirra's necropsy that said she had *not* been shot. No other necropsies she'd read featured such a message. The Parks personnel who compile the necropsy reports about deceased dingoes were probably aware that Parkhurst and SFID obtain copies of the necropsies through Right to Information legislation. I understood that Parkhurst was suspicious that rangers may have shot Kirra and that she thought that QPWS might have been covering their tracks by including a note on Kirra's necropsy, or that, maybe, the note was a message especially for her.

Parkhurst had called Kirra's mate Ronan, which means 'masterless warrior', because he hated people. She described how he always ducked away from people. She thought QPWS might have killed Ronan when they killed the Hook Point pups. If they did, there is no record of it in the QPWS *Humane destruction database*.[1]

Now I wish I could remember more vividly the thick bush, the palms and eucalypts of Coolooloi. Parkhurst told me that there had been a dingo pad a few hundred metres from the track. They rendezvous at different sites, or pads, at different time of the day. I so badly didn't want to miss anything that I had trouble concentrating. I was 'stupid with too many experiences'.[2] Even though she was showing me, at that stage I did not fully comprehend how much joy and pain Coolooloi held. She told me that in 2009 a hazard-reduction burn had got out of hand and 20,000 hectares at the southern end of the island had been badly burnt. She believed that QPWS burnt on purpose at whelping time to wipe out pups in their dens. She told me how three female dingoes and one male dingo rescued one pup each from three different packs and cooperated to feed and raise them. At the Coolooloi

---

1   Allen et al. 2015, supplementary material
2   White 1963, 11

Creek camping zone, where we did not get out of the car, Parkhurst turned the LandCruiser around and we headed east. It took about half an hour to travel eleven kilometres along the bumpy inland track that made Parkhurst feel motion sick before we emerged onto the eastern beach, which, by contrast, was serene and quiet to drive on – almost like sailing.

Parkhurst drove; we talked and watched for dingoes. Burnt stumps of trees lying on the sand looked like dingoes, but there are no black and tan dingoes on K'gari. The visit to Cooloooloi must have brought back some difficult memories. Parkhurst told me that QPWS killed Kirra's pups because, according to the rangers, they had become aggressive but, she said, the pups didn't bite the rangers even when they were trying to kill them. The killing of the Hook Point pack was not particularly unusual. In 2009, ten dingoes, including the Hook Point pack, were killed by QPWS; between 2001 and 2013 QPWS staff killed an average of nine dingoes a year on K'gari[3] with firearms or by trapping and sedating them before they administered a barbiturate overdose by injection to the heart.[4] Through a Right to Information request Parkhurst found out that rangers did not follow the usual practice of anaesthetising the four young Hook Point dingoes before they killed them with an injection of Valabarb (pentobarbitone sodium) to the heart. She claimed there had been no incidents for her dingoes before a researcher called Rob Appleby started collaring them, which scattered the pack. 'They came to the beach,' she said, 'and got hazed mercilessly.' Another year, she told me, Ronan killed some of his pups who had been tagged by QPWS. She said she had begged them not to tag her dingoes.

She described how each year the pups were born and for three months, during the wonderment of the puppy season, they were fed by their mothers. 'They don't know what they've got coming,' she said, 'year after year. Who will survive?'

We arrived at Eurong, hub for the biggest industry on the island – accommodation and food services. Facilities included a bakery, a general store, a petrol station, a pub, a bottle shop and the Eurong Beach Resort. There are other human settlements on K'gari – Kingfisher

---

3   Allen et al. 2015
4   Ecosure 2012, 81

Bay Resort Village, twenty-two kilometres away on the west coast, and Happy Valley and Orchid Beach, twenty and sixty-six kilometres further north, respectively, on the east coast – but Eurong is the most populous, home to seventy-eight of the island's 194 permanent residents, according to census figures from 2006 and 2011. Coincidentally, the estimated number of dingoes on the island – between 76 and 171 – is similar to the number of human residents.[5] It is probably no coincidence that Eurong is also an important dingo population centre.

Before Eurong and the neighbouring village of Second Valley were enclosed by a fence designed to keep dingoes away from human habitation in 2008,[6] dingoes and people lived in close proximity. 'The residents knew all of the different generations of dingoes,' Parkhurst said. 'There was a huge cohesive pack at Eurong. It had about fifteen and it was multigenerational. I think it was five generations in this pack'.

I could picture dingoes lying around one of the resort's pools, delicately dipping a white paw in the water. But, despite the village's laidback, holiday feel, between 2001 and 2013 Eurong dingoes were 'subjected to destruction over multiple successive years'[7] with 'one to four dingoes … humanely destroyed … each year since January 2001 (except 2005 and 2012, when none were destroyed)'.[8] Of the 76 dingoes killed by QPWS between January 2002 and October 2012, twenty-four of them were killed around Eurong and Dilli Village, eleven kilometres south.[9]

We drove across an electrified cattle grid to enter the town. It was meant to keep the dingoes out but the grid filled up with sand, Parkhurst pointed out, and dingoes tiptoed across.

'The locals knew when the dingoes whelped their pups,' she explained. 'The dingoes used to whelp their pups in the townships a lot of the time and get the people to look after them while they went

5   Allen et al. 2015
6   Field 2008
7   Allen et al. 2015, 14
8   Allen et al. 2015, 15
9   Ecosure 2012, 84

hunting. And the people in the townships were able to keep an eye on what was going on with the dingo population.'

I imagined dingoes dozing in the shade on verandas; dingoes watching people from a cool, safe place under the houses; people chatting to dingoes as they went about their business.

'So they fenced them,' Parkhurst continued, 'in effect gagging them, in effect separating them from the dingoes. And a lot of dingoes died when the fences went up 'cause they pined to death and they weren't able to get to their people anymore. And yes … they got fed in the townships. But basically it was more about lounging around in someone's front yard and being with their people. They loved their people.'

We sat on the dune up the beach from Eurong, among dingo tracks in the sand, and ate the rolls Parkhurst had made for us. She described what a quick, clean hunter Kirra had been. Her pups hadn't liked snake or goanna much, but there were wallabies and bandicoots and echidnas. She explained how the island dingoes are omnivores and eke out an existence on berries, nuts, fruit, grasshoppers, foliage, whatever.

Tiny opalescent welcome swallows swooped and flitted, copper and indigo, around the LandCruiser, our sand-ship. When Parkhurst told me that she would have died if she'd had to serve time in a correctional centre because it would have been easy for the prison officers to withhold finger-prick testing equipment and sugar if she had a hypo (low blood glucose) and explain her death in some other way, I believed her.

My brother Roderick was jailed twice in Queensland in the late 1980s. Before he was jailed the first time, in early 1987, he rang home many times in an agitated state to tell Mum how Coolangatta cops had shot his cattle dog Alex (one of Possum's pups) and stolen his buffalo-wool jumper, which Mum had brought home to him as a gift from Canada. I wrote in my diary that he doused a lawn in petrol – I didn't record whether he lit it. Brisbane's *Sunday Mail* reported how, on 22 February, with a towel wrapped around his right arm, he ran into a dive shop as the owner, Mr Grayson, was taking the float out of the cash box and putting it into the till. My brother pointed his arm at Grayson and said, 'There's a .38 under here. Give me the money out of the till.'

Grayson replied, 'Oh fuck off.'

Rod said, 'I'm serious, you know. I'm not scared of dying.'

Grayson said, 'Oh look, fuck off, will you. Get out of here.'

My brother put his arm down and said, 'Well, what's the time then?'

Grayson told him it was ten to nine and Rod ran off toward the Broadwater. He went back to the dive shop later that morning and Grayson asked him what he wanted this time before telling him, 'If I was you I would go because the police are looking for you.'

Roderick replied, 'Oh yeah, they want me to give a statement on corruption.'[10]

He was placed in a forensic, or criminal, psychiatric ward at Wacol Prison. He was not allowed to put up the postcards we sent him on the walls of his cell. He read Roald Dahl stories and modelled clay heads and busts of women – his three sisters, he said.

When my sister, his best friend and I visited him at Easter we sat at a bench in a room with high brick walls, and chatted and played cards. He wore a royal blue prison uniform, drank a lot of the Coke we'd brought, and ate the hot cross buns and chocolate. I remember him smoking a lot of cigarettes. He wouldn't miss the sixty-six other inmates, he said. One of them had fucked a duck. He walked erect, with his chest out, and waved as he passed through a wire enclosure on his way back to the cell block.

Rod had started going to Queensland in the early 1980s on what, at the time, people might have called 'benders'. He didn't sleep; he was itinerant, driving around in his ute with his dog – till he crashed his car and lost his dog. I'm not sure when we started to call them 'breakdowns' – perhaps when we started to realise that, even though he appeared to embrace this agitated energy, these episodes were not voluntary. In May 1987 the Queensland Mental Health Tribunal found him to have a defence of insanity so he did not have to stand trial on attempted robbery.[11] Rod didn't think there was anything wrong with him but the patterns and relationships he saw everywhere must have both stimulated and exhausted him. He brought what he knew of New South Wales crime with him to the Gold Coast and decided that the prostitutes working at the then-new Jupiters Casino were connected to the police. I can see him, barefoot in his filthy trench coat, his hair matted, his two front teeth long gone, loitering in the foyer at

---

10   Allen 1987
11   Allen 1987

Jupiters. I had been angry with him, and afraid of him, but I also knew he was vulnerable. When he was well he was a cooperative, considerate, courteous person. The repression and inconsistency of institutionalisation did not help him.

In 1989 he went back to Queensland and was detained in a secure psychiatric facility called the John Oxley Memorial Hospital at Wacol. There he was injected with the maximum dose of an antipsychotic drug called Stelazine (trifluoperazine hydrochloride) while our mother again went through the legal processes to bring him home. When he came back to Sydney he was hardly recognisable. The expression on his face was fixed, zombie-like; his brow was drawn down over his eyes in an immobile angry stare. He had put on weight and lost his coordination. Food dribbled out of his mouth and down his chin. His body was rigid but he could not stand still. He moved his weight from foot to foot, rocking a little while he lifted one foot slightly off the ground and put it down again in a movement known as 'the Stelazine shuffle'. The medical staff at Wacol had not reduced the dosage of his medication for the six months he was incarcerated. It could have been neglect. Or ignorance. Or wanton sadism.

I did not tell Parkhurst how her mention of dying in prison made me think about my brother's experience. We walked a little way along the beach and I asked, 'What would you do if a big cross alpha male came bowling up to us now?'

'Bring it on!' she said, as if nothing could make her happier.

After lunch we drove north. The eastern beach is long. It feels endless. On our right were the constantly breaking waves, in front of us the wide expanse of wet sand, and to our left the rise of the dune at the back of the beach with its straggly casuarinas. Occasionally we crossed a creek that wound its way around a low part of the dune or carved its way through the foredune and gushed across the beach to the sea or dispersed on the sand.

The dingo packs Parkhurst told me about defined the seemingly monotonous terrain, gave it meaning. She pointed out the territory of the Cornwells pack and the neighbouring One Tree Rocks pack. The One Tree dingoes and the Cornwells dingoes shared Lake Wabby, she said. She told me how the One Tree pups used to come out at 6 a.m. and greet her, jumping around, beside themselves with joy. 'I don't know why they like us,' she mused. 'Maybe we were nice to them once.'

North of the Gabala camping zone, I chatted about how I met my husband. My mother and I were at a Sydney Festival production at the Wharf Theatre when we ran into some friends of mine, and he, just returned from two years working overseas, was with them. I don't remember much about the play we were all, coincidentally, going to see except it was about incest and Pia Miranda was onstage for a lot of it in her underwear. There was no allocated seating in the theatre. We took cushions in, chose our own seats and I sat next to this new friend of my friends.

Parkhurst and I were near Poyungan Rocks when she spotted a dingo. She looked through her telephoto lens. The tag was in the left ear, which meant he was a male – females are tagged in their right ear. He was strolling south along the beach before he stopped on a grassy sand rise in the sparse shade of a casuarina and lay down to watch some men fishing about 100 metres away.

'He's all right if he stays there,' Parkhurst said. She estimated he was about three years old. Through the zoom lens she could see a scar on his face.

We needed to get back to the barge at Hook Point so we turned around and left him, a red flash of sunlight on the pale beige sand. An eagle soared in the sky in front of us as we drove south. It was about 3.15 p.m. Since we had arrived on the island at about 10.30 a.m. Parkhurst had counted five QPWS vehicles.

As we drove south I told Parkhurst about how my husband and I had first got to know each other. After we had gone out in a big group for St Patrick's Day, he had given my friends a Lonely Planet travel guide for China to pass on to me because I was about to visit a friend in Shanghai. All through that trip I anticipated, with great delight, returning his book to him so I could see him again. I don't know that Parkhurst was particularly interested in this story but because talking about my husband had brought us the good luck of seeing one dingo, I continued. South of Eurong, I broke off when we saw two dingoes jaunting nose to tail, etched liked cave carvings, along the eroded side of the sand dune at the back of the beach. They were both tagged in their left ears. Parkhurst assumed they were brothers, about ten months old.

They moved from the dune to the swash zone and back. One pulled at the roots of some dune grass with his mouth and the other trotted out across the beach toward a man who was looking for pipis or worms in the sand below the high-water line. The dingo sniffed the sand where the man had been, watched him and followed him at a distance of a few metres. The man took no notice. As the young dingo stood calmly by the man's vehicle, we nicknamed him Bold.

'He's too social to survive,' Parkhurst said.

After the man drove off, Bold lay on the sand and rolled. His brother came over to him, sniffed the area where the man had been, and they continued travelling south in front of the dune. Bold's white-tipped tail was crooked. His brother, who was slighter, limped. In the distance I saw the lanky frame of one of the dingoes squatting slightly, his nose pointed straight at us, while he defecated.

We drove slowly at a distance from them as they travelled down the beach. Backlit in the afternoon sun, Bold strode along the top of the dune. His tail was curled over his back; his ears were forward. He waited for his brother and, cheek beside flank with an easy companionship, they passed the Gerowweea Creek sign. The smaller brother jumped down from the dune first, a graceful leap full of air, and waded chest-deep into the creek to drink.

Bold followed and on the other side of the creek they played. They ran and rolled and somersaulted. Bold bowled his brother over. They leapt and sparred with their jaws and ran some more. They mouthed and pawed and lay low and sprang. According to canid ethologists, juvenile play is a 'collage' of the neonatal behaviours young canids are growing out of and the adult behaviours they are growing into.[12] When young canids play they are creating new neural tissue, growing densely connected brains that will help them become adults.

Afterwards they arced back north out on the wet sand where, with apparent calm, they investigated strands of pale brown sea foam. They looked as if they were searching for something, I thought. Their bodies glowed saffron in the afternoon sun and cast motile golden reflections on the luminous sand.

---

12 Coppinger and Feinstein 2015, 176

Then they trotted in front of another car and headed south before, at 3.37 p.m. they disappeared inland into the scrub near Govi camping zone. Parkhurst stopped the car and got out. While she rolled a cigarette, I took photographs of the dingoes' paw prints in the sand.

I was about ten metres away from the vehicle when Bold and his brother emerged from the bush. Bold came trotting down to me; his brother must have gone somewhere else on the dune.

I stood there. He pointed his nose toward me. *What will happen here?* I thought. I was aware of blood pounding at the back of my head, a cold sweat on my skin.

I am not here to dominate you, I thought. I tried to let him take his own course of action. I didn't know him well enough and I was not confident enough of him or the situation to greet him as required by dingo etiquette, squatting down, touching noses with him. I tried to talk to him gently with my eyes, which had nothing to give him but interest, possibly fear, and firmly with my body, which was too old, too rigid, too full of human stories to play with him. I saw his multicoloured ear tag as a pattern, like the shape of his eyes in his arrow-shaped head.

Parkhurst took fifteen photographs during the one minute and four seconds that our encounter lasted. They reveal a sequence of movements that I did not perceive at the time. She captured his approach, step by big-headed, ginger step:

He changed direction slightly to come from downwind.

The moment I perceived as his flinching, his left front leg extended toward me, his gaze down at my knee occurred early in the sequence.

I tried to look nonchalant while he walked clockwise behind me.

He stopped behind my left leg to look at my face.

Later Parkhurst told me she thought he might have been deciding whether or not to mouth my leg. Perhaps I had not read him accurately but it did not occur to me, at that stage, that he might nip.

I wasn't going to run. I don't know why the freeze response is not talked about as much as the fight-or-flight response when animals, including humans, face situations that might become threatening.

He waited. I didn't do anything. So he strode anticlockwise to face me front-on again.

We looked at each other. I was almost vertical, leaning slightly toward him. He was horizontal, addressing me. I didn't take a photo of his face – so close, inquiring – because I didn't want to excite him so I didn't move my hands to raise my phone with its camera to my face. I didn't want to bring attention to an item he may have been interested in. I didn't want to take my eyes off him to put a contraption between my eyes and his eyes. And maybe I was too scared to move.

I had straightened up – my elbows were close to my sides, my gaze downward – when he did what might have been the beginning of a play bow, a last split-second attempt to get me to do something interesting. The photograph shows his left front paw slightly off the ground, his tail swishing, his jaws very slightly open, his eyes on my face. His invitation to play must have lasted just a fraction of a second. I don't remember it.

I was boring and he walked away from me toward Parkhurst with his tail in the air. He looks light on his feet and his triangular eyes give me no clear indication of what he is thinking. His posture looks proud; his expression looks aloof, maybe. Without any more incident he circled around and headed south. I was grateful to his brother, who had lain low and stayed out of sight somewhere on the dune.

Afterwards these moments resonated beyond the short duration of our meeting. What happened? Maybe he sensed how afraid I was. But to my mind he was civil – in a dingo way: he came to see who I was; he addressed me with his face; he sniffed me; he left of his own accord. In retrospect I realised that every fraction of a second held all sorts of divergent possibilities. My fear was involuntary but my conscious mind knew I must behave calmly, I must appear to be unfazed.

Perhaps the frightening dingo education contributed to my fear, as well as a primal sense that a dingo's jaws are large and powerful. As Parkhurst pointed out, if a dingo wanted to seriously injure or kill a person, they could. Even though I knew Parkhurst was close by and ultimately I was safe, I was relying on Bold's choices too, his – for want of better words – instinct, good sense. I don't think he tried to threaten or intimidate or show aggression toward me. He seemed curious. Even though I was afraid I am glad I met him.

My encounter with Bold was typical in some respects. Between 2001 and 2015 male dingoes were involved in most reported incidents, and male subadults, between one and two years old, a few months older than Bold was when we met, were the highest reported sex-by-age category.[13] Where information about location was recorded, almost two-thirds of serious, Code E incidents took place on or close to the beach, not at campsites, lakes, townships, resorts, on boardwalks or in the bush.[14] Of the 160 serious incidents recorded over this fifteen-year period, more than a quarter (forty-three) occurred at Eurong.[15] But 'dingoes rarely seriously injure even the most vulnerable of people', 'nor do they regularly seriously injure people more generally'.[16]

That night, back on the mainland at Debbie's Place in Rainbow Beach, I ate lentils and rice left over from the Indian take-away I'd bought the night before. I didn't want to read scientific articles about dingoes on Fraser Island. I didn't want to think about Parkhurst's polemic about culling and dingo management. Sadness welled up in me and stayed like big seas and king tides leaving pools on the Spit on the southern tip of the island where the dingoes go fishing.

Would he be treated with the respect that he showed me?

What would matter to Bold? Not how bold he was, or how doomed, or how much his brother would miss him if the rangers killed him. They were so attuned to each other that afternoon, out for some fun, out for a feed, ranging over the beach, waiting for each other, regrouping, playing, watching each other – their closeness just part of beautiful life. What would matter to Bold, I thought, was food. I got a feed that night. I wondered whether those brothers did.

The next morning I was still thinking about my meeting with Bold. How would I be able to visit K'gari with my family? With my daughter? She was eight at the time and nervous around strange dogs, let alone free-wheeling dingoes. I needed to talk with a QPWS ranger. How could they manage a place where the dingoes were so free and confident? How could they stop dingoes from mingling with people?

---

13 Appleby et al. 2017, G
14 Appleby et al. 2017, E
15 Appleby et al. 2017, D
16 Appleby et al. 2017, H

I took only one wrong turn on the way back to Hervey Bay, which took me on a detour to the fishing port of Tin Can Bay. I arrived back in Hervey Bay in the late afternoon and checked into the Mantra. My room overlooked the marina, drenched in golden light, and K'gari, olive and lilac under a pale pink sky with violet clouds. I rang Naomi Stapleton, the QPWS conservation officer who had been communicating with me about interviewing rangers, to thank her for being so prompt in her responses and to ask whether there might be a better time, when QPWS weren't so committed to other research projects, for me to request to interview rangers. I wanted to demonstrate that I was a reasonable person, to tell her how valuable rangers' first-hand perspectives would be, to reassure her that it would be only an hour or so of a ranger's time.

She was helpful and friendly. She directed me to the Queensland government Department of Environment and Heritage Protection website, which listed the research-project applications that had been successful under the Fraser Island Dingo Conservation and Risk Management Strategy. Other government websites that provided information about dingo research were the Department of Science, Innovation and Technology and the Department of National Parks, Sport and Racing. She suggested I talk with dingo scientist Ben Allen and refer to the Ecosure report.[17] She seemed a little nonplussed that so many dingo researchers wanted to work on Fraser Island and not somewhere else – like central Australia. I was enthusiastic about how unique K'gari is, how people can see dingoes so easily, as they can't disappear into the desert like they do in central Australia. She told me that there are some packs on the island that no one ever sees. 'But the dingoes on the eastern beach,' she said wearily, 'they've been bred on the eastern beach and they'll breed again and they'll breed again ... '

She suggested I try again, later in the year.

That evening, after I had returned the little pink rental car, before dinner, while I was bending over some maps spread out on the bed, my back went. Only it didn't go anywhere. It stayed with me and decided to give up being a reliable, helpful back. It hurt and I couldn't move. In the middle of the night a drunk man shouted and banged on the door

---

17   Ecosure 2012

of the neighbouring room. 'Let me in, ya bitch.' Bang bang bang. Too close. 'I'll get in. I'll show you.' Bang bang bang. I lay petrified.

Early the next morning I knew I was not going to be able to take the tour of the island that I had booked. Very slowly I did all the things I had to do – rearranging travel and accommodation and an interview, asking for help with my luggage, checking in for my flight, waiting in departure lounges, sitting on planes – before my husband and daughter met me at Sydney Airport and took me home, where for a couple of weeks I mostly lay on my stomach and read about K'gari while my back slowly recovered.

* * *

On 17 August 2015 the Queensland Department of National Parks, Sport and Racing issued a press release reporting that a male dingo had been 'humanely destroyed' on K'gari after a number of aggressive incidents. The dingo was killed after he bit a nineteen-year-old male tourist on the thigh on the beach in front of Eurong on the morning of Sunday 16 August. He had also bitten a woman just north of Dilli Village. In mid July, after QPWS rangers were forced to close the beach after his involvement in a number of threatening encounters, he had been fitted with a satellite tracking collar so that rangers could monitor his movements and respond quickly to possible human interactions.[18]

I emailed Parkhurst and Save Fraser Island Dingoes (SFID) to see if they knew any more about this dingo. I hoped it wasn't Bold. At first Parkhurst thought that the place we had seen him, near Dilli Village about fifteen kilometres south of Eurong, was too far away from Eurong for it to have been the same dingo. She thought there was another pack between Eurong and Dilli. Then SFID found out that the dingo who had been killed was tagged with a purple, green and yellow ear tag. QPWS identified him as PuGY14m for the colours of his ear tag and the fact that he was a male born in 2014.

Even though Bold and I had stood so close to each other on the beach in May and I had noticed his ear tag, I could not have said what its colours were. I zoomed in on my photographs of him. His ear tag

---

18   Queensland government 2015

looked yellow, green and red to me and I thought there were numbers on it, which I could not read.

Parkhurst explained to me that colours, not numbers, were used to identify the dingoes and that sometimes the ear tags of dead dingoes were reused. After a week or so of checking, Parkhurst emailed me confirming the dingo QPWS staff had killed on 16 August was Bold. She must have checked her own photographs. We were both sad but Parkhurst was frustrated and angry, too, at how 'repetitive, predictable and avoidable' Bold's death was.[19]

Later, over the telephone, we discussed her photographs of Bold and his brother. 'They always have to patrol the beach every day,' she explained, 'because it's part of their territory. Some dingoes patrol the beach at low tide, some at morning and night. It just depends on the dingo when they're going to do their patrol, and while they're doing that they have a play.'[20]

Parkhurst did not believe a dingo would attack a person for no reason. She disagreed vehemently with the QPWS view that when a dingo lowers the front part of its body in a 'play bow' they are dominance testing. She interpreted that behaviour as an invitation to play, an act of curiosity and sociability. 'They know we're a different species,' she said, 'but they don't know our skin isn't fur and if they bite it it's going to puncture it. They don't know we don't understand them. They just assume, "Here's someone to play with."'[21] It occurred to me that we humans do not realise that we do not understand dingoes, either.

In a new communicative spirit, Neil Cambourn, executive director of QPWS regional operations east, had been liaising with SFID about Bold. He had video footage of Bold's erratic behaviour, which he offered to show at a SFID meeting. I looked forward to seeing it and, with plans to return to the Fraser Coast, I wrote again to QPWS requesting to interview rangers. This time I emailed Ross Belcher, Fraser Coast principal ranger, and pointed out that, even though I was intending to write a book, my first objective was to produce a doctoral thesis, which would be confidential and not available publicly. I was astonished

---

19   Parkhurst 2015c
20   Parkhurst 2015d
21   Parkhurst 2015d

in a very good way when Ross Belcher replied with the names of three rangers whom QPWS was happy for me to interview with the stipulation that I provide QPWS and the Butchulla Prescribed Body Corporate Board with a final draft of the interviews before I included them in my thesis. Subsequently, I asked them for permission to include their interviews in this book.[22] I wrote to the rangers to introduce myself and arrange a time to meet. I also telephoned Butchulla elder Kay Skelly[23] to ask whether she would let me interview her. Her number had been given to me by a contact of Parkhurst's.

In mid October, around the second new moon since Bold was killed, I started to read the QPWS dingo interaction reports for 2015, obtained by SFID through Right to Information requests and shared with me by Karin Kilpatrick at SFID. The reports are full of funny anecdotes and information about dingoes' culinary preferences. For example, dingoes rearrange tea towels at campsites; they eat all the peanut M&Ms from a large box they find in a tent and leave the dried noodles from a packet they rip open largely untouched. But the reports also turn dingoes like Bold, who are sociable and curious, into delinquents.[24] They describe typical juvenile dingo behaviour in old-fashioned terms used to describe human criminal activity. If dingoes find themselves in places frequented by humans, which is not hard when the camping zones and tourist attractions in their territories fill up with people at Christmas, Easter and school holidays, they are 'loitering'. If dingoes are interested in human food, they are 'soliciting'. If dingoes make away with anything from a human camping zone, they are 'stealing'.

If humans approach, lure, feed and hit dingoes, or throw things and drive at them – all actions recorded in the interaction reports – dingoes are culpable. A man was observed kneeling down on the beach south of Poyungan Rocks, coaxing a dingo – possibly Bold when he was eight months old – toward him, then grabbing the dingo by the scruff of the

---

22 The Butculla Prescribed Body Corporate and QPWS personnel Linda Behrendorff, Ross Belcher, Finn Dwyer (not his real name) and Dan Novak have read and reviewed chapters based on their interviews. Jennifer Parkhurst and Kay Skelly (not her real name) have read and reviewed chapters based on their interviews.
23 A pseudonym has been used here and subsequently.
24 Foucault 1995, 277

neck with his right hand, slapping the dingo in the face and pushing it away. There were no repercussions for the man and it is only because an onlooker took the trouble to report the incident that it is recorded. But the dingo, who retreated to the dunes, received a Code C report for loitering at a recognised visitor site (people nearby), soliciting food and being fed or encouraged. If a human gives a dingo a steak, the dingo receives an incident report.[25]

Bold accrued many reports after I met him. The process for the creation of the interaction reports warrants its own examination but I was reading them for the story of Bold. I wanted to make chronological sense of the last three months of his life. I wanted to work out what had happened to him. It was hard to digest all his adventures. He burnt bright. He lived large.

---

25 Accounts of human–dingo incidents are taken from QPWS interaction reports sent to me by SFID. See Queensland government 2017–2018.

5
Let's dance

All was music. The picking up and setting down of their feet, certain
anglings of the head, their running and resting, the positions they
took relative to one another
      Franz Kafka, *Investigations of a Dog*, 1922 [2017]

'Al-loe-w!' said the recorded voice of a young man when I rang the
tour company's toll-free number to re-book the two-day tour of K'gari I
had been unable to take in May. 'I am the bay-care from Eurong Beach
Resort. Nice to 'av you visit to this side of zee island. Ha ha ha ha ha!
'Ere we're only a 'op, skip and jumper away from the mainlan-d. It often
feels like another planet. We speak of it as zee call of zee wilder ...' He
went on – about swimming pools, restaurants, bars, the general store,
the 'lovelies' he was just taking out of the oven at his bakery, lakes,
four-wheel driving, beach fishing, 'truffles of your own' and 'umpback
whales. The recorded *bay-care* passed me on to a recorded fair dinkum
bush-tucker ranger, who passed me on, via the sound of a gamelan
orchestra, to a recorded throaty girl-woman 'therapist' who was 'in the
middle of performing one of our specialised island massages. No peeky
weekies now!'

    'So,' she said, 'while you're on hold, let me help to release that
on-hold tension you might be experiencing. Imagine me massaging
your tired aching limbs with a blend of our natural island products. Feel

the oils as they caress deep into your aching muscles ... Feel the tension float away like a lotus petal on a crystal mountain stream ... You are deeply relaxed ... Your body goes limp. You drift into another world. Your arms fall. Ah, you've dropped your phone. Ha ha ha ha ha.'

'You are at position five in this queue,' another voice interrupted. 'Please wait to be connected.' But the throaty therapist was not gone, reassuring me with what became a mantra of the recorded message: 'No worries. Our consultant won't be long now. While you're waiting, here's a friend of mine ...'

The company that owned this tour company also owned the Kingfisher Bay Resort, the Eurong Beach Resort, Fraser Island Ferries, a golf course on the mainland and Cool Dingo Tours. Unfortunately I was too old to take a Cool Dingo Tour. Running ninety per cent of the accommodation options and the majority of touring experiences on the island[1] must have kept them busy because the recorded message continued for more than eleven minutes. There was a ferry captain. A kookaburra laughed. Bubbling noises, apparently made by a humpback whale, were interpreted and the refrain 'It's a Fraser Island thing. No worries' was repeated with no indication from the speakers that they had any idea of the strange impression their recorded message made.

* * *

This time, I packed light – no candles, no megaboom speaker, no paper articles, only one printed book, *How Dogs Work*, written by two ethologists, Raymond Coppinger and Mark Feinstein, and published in 2015. The plane flew in from the ocean over the distinctive straight line of K'gari's long eastern beach and descended through a blanket of cloud that had been hiding oyster-coloured lakes and the white cliffs on the west coast of the island. The azure straits and their low-slung islands came closer and closer until I could pick out details of mangroves, and tin roofs and car bodies in Hervey Bay.

I was glad I had selected a rental car from the second cheapest instead of the cheapest category for this trip. I had a different room at the Mantra, which was small and comfortable, tucked in over the

---

1   Schlesinger 2018

marina with a balcony that had a view toward the island and a massive, noisy air-conditioning unit. The official temperature was thirty degrees but it felt hotter. The water out of the tap did not taste good. I emptied the jug in the fridge of its cold water (who knows what might be in there), boiled some new water, put it in the jug and put the jug back in the little bar fridge.

I had organised an interview with Kay Skelly,[2] a Butchulla elder, for that afternoon. She rang me from the Air Fraser terminal to say her plane from the island had landed. She told me she was with Rosie, an English backpacker who had been staying at K'gari camp and whom Skelly had taken under her wing. I told her I would come to pick them up straight away but not to worry if I took longer than she expected because I might get lost on the way. 'We'll send a black tracker out for you,' she said.

A few weeks earlier I had telephoned Skelly from Sydney and, early in our conversation, she half-apologised for swearing. I told her I swore too but my children didn't like it. 'I swear because I get so angry,' she said. 'I swear at the rangers.'

Skelly explained how she spent some of her time on the island at K'gari camp and some in Hervey Bay. The K'gari Educational and Cultural Centre, a camping area owned and run by Butchulla people, was, she told me, located between the rusting wreck of the steamer *Maheno* and the coloured sands, on a track up to Lake Allom where, she said, the water is lovely and soft. It was November. She'd just seen a king brown snake on the island and she wasn't sure if she'd go back because there were too many snakes. She was scared of snakes and spiders. 'But I'm not scared of dingoes,' she said, 'not our dingoes.'

During that first telephone conversation Skelly talked about some of the K'gari dingoes she knew: Brian and Shirley; Sister Girl, Pup Pup and Narawi, whose name means white waves on the ocean; and Inky and Winky. 'Inky Winky' she said, like sand falling gently, fluidly, through a funnel.

I had heard about Inky. He was born around July 2011 and was recognisable because his ear had been unskilfully tagged and hung down with an infected-looking wound.[3] He was the subject of over

2   A pseudonym has been used here and subsequently.
3   Bryant 2012

20 incident reports and, according to newspaper articles, QPWS killed him because of four Code E incidents, which included his lunging at people and their children, his running out of the bush at people playing volleyball on the beach, and his grabbing hold of at least two tourists with his mouth, though he did not break their skin.[4]

I deduce from the Queensland Parks and Wildlife Service's *Humane destruction database* and Save Fraser Island Dingoes' press releases that Inky's brother Winky was killed by QPWS staff at North Woralie Road near K'gari camp on 7 August 2012. Another brother, Byron, who had received five Code C and five Code D incident reports but was never recorded as having nipped, bitten or acted aggressively,[5] was trapped and killed on 4 October at Cathedral Beach. Inky was finally trapped and shot at the Dundubara bin compound on 22 November.[6]

I found Skelly and Rosie waiting on the road to the airport out the front of the Air Fraser terminal. Skelly asked me to stop at a pharmacy on the way back to town so she could buy some hair dye. Back at the Mantra I offered them fruit and tea or wine. Skelly didn't drink and she was not impressed with the warm water I gave her from the fridge. Rosie sat on the balcony while we talked.

Skelly didn't want me to use her real name. She didn't want to jeopardise the chances of her nieces or nephews if they tried to get a job with QPWS. So, later, to make the task of giving her a pseudonym pleasurable, I looked through lists of Butchulla words and Gaelic words. There are many beautiful Butchulla words – *djaa* meaning 'earth', 'land', 'dirt', 'ground', 'place', 'town', 'country', 'home'; *dji'nang* meaning 'foot', 'paw' or 'claw'; *dji'nang-djaa* meaning 'track', 'path of foot', 'mark on ground'; and *dji'nang'gou* or 'on foot'[7] – but I didn't want to choose the wrong one, so I chose the name Skelly, from a Scots word *skellie*, which is a 'ridge of rock on a seashore covered at high water'.[8] It is a mellifluous word, and denotes something hardy and able to withstand

---

4   Walker 2013
5   Bryant 2013
6   Allen et al. 2015, supplementary material; Anon. 2012
7   Bell and Seed 1994, 25, 29
8   Grant, n.d.

tides, waves and storms, something sometimes submerged, something solid and under, and possibly subversive.

One of Skelly's eyes drooped a little at the start of our interview, from being tired, I thought. But her face was animated, lively, when she talked about how, before QPWS put fences up around K'gari camp in 2012, Inky and Winky used to lie on the veranda of the house with their heads resting on each other while they watched TV. 'When someone come up to the house, they had to step over them.' We laughed at the idea, the bulk, of those dingoes lounging on the veranda making people step over them. 'They just laid there,' Skelly said.[9]

On a QPWS interaction report dated 15 January 2015 the house at K'gari camp can be seen in the background of a picture of an empty metal bowl. The interaction form reports an unknown dingo for 'loitering at recognised visitor sites (people nearby)' and 'being fed or encouraged'. Two rangers who were inspecting the fence found the bowl. It looks empty in the pictures on the report but it is described as containing cooked mince and vegetable residue. The offending bowl is on a sandy grassy patch just outside a person-width, metal-framed chicken-wire gate. The wooden-topped, square chicken-wire fence either side of the gate is just higher than the QPWS Toyota, which is also in the picture. Up a walking track fringed by what looks like thick teatree that grows taller than the fence is a wooden building constructed on wooden stilts that are about one metre high. Narrow stairs on the right-hand side lead up to a veranda enclosed by a wooden paling fence. Before K'gari camp was fenced it would have been a cool, shady, well-frequented place for young dingoes to rest.

I don't know whether it was because I wanted to be friendly, but there was something familiar about Skelly's face. It was easy to listen to her voice. She had a particular sing-song intonation with a one-syllable word like 'Yeah' that started high and descended, like a door opening.

Inky Winky responded when she called them, she said, understood when she said 'Naughty' to them, and the rangers saw what quiet dogs they were. 'They're good dogs,' she told the rangers. 'They don't bother anybody.'

'What happened that Inky and Winky got killed?' I asked.

---

9   Skelly 2015. Subsequent direct quotes from Skelly are from this interview.

'There was a German backpacker,' she said, 'and they got no sense of direction. He goes down to the toilets. Instead of going back the main way, up the track where all the lights are – lights everywhere – he went that way and walked up the track – the wrong way, away from K'gari camp. These other dingoes got him. They got into him.'

'They nipped him?'

'They got into him. He even climbed up a tree. He was that drunk he fell out of the tree. I think they said there were about four or five, a little pack. They did rip him.'

The twenty-three-year-old German backpacker gave his name to journalists as Justin. In the newspaper photographs taken in Hervey Bay Hospital after he was airlifted off the island he has puncture wounds and swollen, torn flesh on his right calf,[10] a bandage around the lacerations on his scalp and tape holding together his lacerated right eyelid, as well as many smaller wounds on his face.[11]

According to the newspaper accounts, on the night of the attack, after drinking a lot, he got up to go to the toilet at around 2.30 a.m., became disoriented, took a wrong turn onto Woralie Road and walked about 800 metres away from K'gari camp before he sat down and went to sleep. Some time later he was startled awake by dingoes sinking their teeth into his limbs. He tried to fight them off, hitting out at them as they bit him. But if he scared one off, another ran in, snapping, biting and taking turns to clamp down on his arms and legs. His screams and cries went unheard. When he played dead the dingoes kept going, moving away and coming back. When they backed off he ran into the bush and climbed two metres high into a tree, but the branch broke. He had to wait until dawn before he could find his way back to camp.[12]

QPWS could not prove that Inky and Winky were there when the backpacker was attacked, Skelly told me. 'It was some other dogs,' she said. 'Might have been their brothers or sisters.' According to Skelly, when the injured backpacker was airlifted out after the attack, Inky and Winky were sitting beside the cameramen filming the rescue helicopter. 'What does that tell you?' Skelly asked me. 'If they were savage they'd be growling at

10   Norton 2012a
11   Norton 2012b
12   Norton 2012a and b

them and trying to bite them. The drivers that bring all the backies in were saying they were here with us because they loved them too.'

'The backpacker didn't want them to be killed or anything,' Skelly told me. 'He said, "I don't want anything to happen to them. It was my fault. I shouldn't have been out there." He took the blame for everything and they wouldn't listen to him.'

Skelly wasn't on the island when Justin was attacked but when she returned she saw rangers in jeans and checked shirts and asked them what was going on. 'How come youse not in your uniform?'

They told her they were looking for a couple of dingoes, one hurt.

She started calling for them. 'Inky! Winky!' But Inky and Winky didn't come. 'The dingoes are cunning,' she explained. 'They know the rangers' uniforms and cars, and they won't come out until they go.'

But she thought something was wrong; she had a funny feeling. When Jim the dingo ranger passed by she called out, 'Come here, I want to ask you something.'

'What?' he said.

'Winky's dead, in't he.'

'Yeah,' he said.

Her granddaughter was cleaning the toilets as Skelly learnt this news. Skelly sang out to her, 'Winky's dead. They killed him.'

Skelly's granddaughter was broken-hearted. 'I hate youse, I hate youse,' she cried at the ranger. 'You're killing my Winky.'

Skelly didn't tell me how many days later she was sitting outside at K'gari camp having a smoke, crying and thinking about 'the dogs', as she called them, when she felt something by her leg. It was Inky. He was injured. 'He had a big hole there,' she told me, gesturing to her neck, 'like someone had put a gun there.'

'Oh my boy,' she said to Inky, 'you've come home. Let me look there.'

He sat and let her look at his wound.

'Why don't you let me help you?' she asked him.

She told me twice about Inky coming up to her with a hole in his neck, once when we talked on the phone and again when we met at the Mantra. It was, I thought, an important interaction that kept playing through her mind. It reminded me of the way I kept thinking about my meeting with Bold. But my meeting with Bold lasted a minute and it was the first time I had ever seen him. Skelly's relationship with Inky was quite different; she had probably known him for the majority of his life.

'I couldn't touch him,' she said. 'Couldn't do anything really.'

I didn't want to interrupt her to ask her why she couldn't touch him – whether it was because he didn't let her – or why she couldn't do anything – whether it was because she could not ask QPWS for help to treat his wound.

'Then he went. They had to bring someone from Brisbane to track him down. They brought a couple up. And they caught him then.'

In four desolate sentences Skelly compressed the last, crowded, hunted months of Inky's short life while he evaded 20 rangers and two trappers specially brought in.

'It's just like losing one of your own,' she said. 'Family. That's how close we get to them.'

She thought that the rangers, who knew how quiet Inky and Winky were, should have put in some good reports about them. There were cameras all over the island and QPWS knew about her relationships with dingoes. But rangers didn't like her giving dingoes names and they told her that she was not supposed to talk to dingoes. 'So I told them straight: "Don't tell me what I can't do. I don't listen to youse. They're our camp dogs. They lived with our people and they never attacked our people because they were their pets."' She blustered. She was frank. She didn't care what Parks told her to do. But she was not self-righteous, or officious, or self-justifying. Her authority belonged to her. The way she said the dingoes' names – Inky Winky, Sister Girl, Narawi, Pup Pup, Bryan and Shirley – was lyrical, intimate, clear.

Skelly traced QPWS's justification for killing Inky and Winky back to the death of Clinton Gage in 2001. 'That's when they started culling them all,' she told me. She said she thought the rangers didn't listen, that money and tourists were more important than dingoes' lives. But she acknowledged that some of the rangers 'love the dogs too', and had to carry out orders to keep their jobs. 'They've got to do what they're told.'

The Federal Court granted native title rights to K'gari to the Butchulla in November 2014, but, according to Skelly, nothing has really changed. 'We've got our native title rights but when we want to help the dingoes we can't do it.'

Skelly told me she didn't think there was enough food on the island for the dingoes. 'I just want to give them a good feed and a big bucket of water.' Rather than transporting all the leftover food from the resorts off the island, Skelly would prefer to see feeding stations set up at various different places for different packs, and tourists could come and see them eating their food. She hated to see dingoes 'starving', as she put it. 'You see their ribs. I don't know how people can see a starving animal,' she said. 'If I was starving my two dogs at home, I'd be thrown in jail, get a big fine.'

In opposition to the foundation premise of the current management strategy and remembering her father's relationship with dingoes at Central Station before the island was managed by QPWS, she told me how she thought feeding dingoes made them less aggressive, not more aggressive: 'They're hungry, poor things. All they wanted was a good feed. They don't worry you so long as their guts is full, bellies are full. They're right. They're happy.'

When I asked Skelly why Parks had the view that K'gari's dingoes needed to be returned to a state of being wild animals, she paused. She looked exhausted. 'I don't know what's wrong with them. They're cracked in the head, I think. Bit wombah. That means silly, stupid. Brundy. That means the same thing, the Butchulla word for silly people. They're brundies. I don't know. Some people say kids are hard-headed. But they're grown-ups and they're worse than kids. Kids'd go and feed a dog, go and pat them and want to be friendly. But these fellas want to kill 'em and shoot 'em. I don't think there'll be any dingoes there in twenty years. They'll kill 'em for any little thing.' She told me she hadn't seen one dingo along the beach that afternoon on her way from K'gari camp to the airstrip at Eurong. Years ago, she said, she would have seen four or five playing on the beach.

'Do you think that's what the strategy is?' I asked. 'Do you think that's what the authorities want to happen?'

'They don't want the dingoes on the island,' she said. ''Cause they want tourists to go over there in peace. You see kids on the beach, little fellas, on their own. Where are the parents? And then the poor dingo'd come along and get shot because they go near them.'

Aware of the threat dingoes might pose to children, she told me that at K'gari camp before the fences, there had not been one dingo attack on a child. The camp dingoes, she explained to me, would protect children, walk them up to the house and down to the fire. Nevertheless she held no hope that QPWS would modify their position and allow people to feed dingoes. She didn't think there would be any dingoes left to feed because of the policy of killing.

'But the tourists want to see the dingoes,' she said. 'When you see a big lot of cars along the beach you know there's one skinny dingo. They're taking photos of it. They run them over, people do. They kill them.' She described how Narawi had a big round lead fishing sinker in her forehead. Skelly didn't know how it got there but assumed a person had thrown it at her.

I asked her who owned the dingoes. 'We do,' she replied. 'S'posed to.' She wanted to care for them, not kill them. 'They know. When we had 'em we loved them and showed our love to them dingoes. They always seem to be friendly with us Aboriginal people, you know.'

Skelly told me a story about how her uncles used to take fishing parties out in the strait near K'gari. One day when the boat broke down, one of Skelly's uncles left the anglers and her other uncle and rowed to the mainland to get the part for the boat. 'When he returned the other old fella was missing. There was some funny business there.' She thought something suspicious, underhanded, had happened.

Skelly's uncle asked the two remaining white men where the other Aboriginal man was. They said they didn't know. They told him they thought he had fallen overboard. When Skelly's uncle went to look for him he saw a big white dingo on the sand dune staring straight at him. The dingo walked and looked as if to say, 'Come with me. Follow me.' When Skelly's uncle followed, the dingo took him straight to the body.

'The white dingoes,' she explained, 'are very spiritual to us ... There's something there. There's something with them dingoes and the Aboriginal people. There is something there that they know about us.'

Butchulla people who worked for QPWS as rangers were not allowed to kill dingoes, Skelly said. 'It goes from the elders to them. We told them, "You're not allowed to be involved in killing them." And so they don't.'

She had never felt afraid of a dingo, but she advised that you should never show your fear. She ignored the dingo that was prancing around her as she was walking back from the *Maheno* one morning. 'I just took no notice and she was following me and she knew I wasn't frightened. I kept looking back and she left me alone. When you run, that's it.' When they pranced around, exhibiting the behaviour that QPWS incident reports describe as 'dominance testing', Skelly interpreted it as playing. She told me she joined in sometimes: 'Like we're dancing together.'

'They know us,' Skelly said. 'They have a routine. They'll do their daily thing. They'll go way down the other end of the beach but they always come back, and that was their routine every day. Managers up there at Cathedral would say, "We've seen Narawi and Pup coming through," and then they'd come home.'

Pup Pup was born at K'gari camp. When one of Skelly's mates didn't know what to call him, she used to say, 'Here pup, come on puppy,' and the name stuck. Pup Pup brought Narawi, who was white, back and according to Skelly they were K'gari camp dingoes for fifteen or twenty years after that.

Skelly told me she didn't see dingoes as wild animals. She thought they were all capable of being just like domestic dogs. But the comings and goings of her free-ranging dingoes were not exactly like the relationships urban people have with domestic pets. Skelly might not see one of her dingoes for a few months. Although it broke her heart when dingoes were killed by humans, she was more accepting of dingo-on-dingo violence. Dingoes were aggressive to other dingoes, she said, when another dingo came onto their territory. 'They know whose territory is whose because they pee around and leave that smell there. That's their territory.' She had seen Pup Pup run out Boy Boy who was hanging around one minute and gone the next with Pup Pup in hot pursuit.

Skelly recounted to me the last time she had seen Pup Pup. 'It was mating time and he had another little chick and I said, "Pup, what are you doing with her? You're married to Narawi. Where's Narawi?" and he'd look at me and he'd look up at Cathedral Beach. Up there she was.

'They point,' she explained. 'They show which direction by looking.'

Pup Pup and Narawi got 'kicked out' a couple of years ago, according to Skelly, when they were too old to fight and 'gave in' to the juveniles coming up from Happy Valley and Moon Point looking for territory. 'The juveniles, they call them, are the ones that want to take over places. They'll come and then they'll take over. The dogs go to another place, and the juveniles will probably go there and kick them out and the other ones'll sneak back.'

Skelly described some of the dingoes to me. After Pup Pup and Narawi, there was Sister Girl, Boy Boy and Bra Bra. More recently, Brian and Shirley lived nearby and visited. Like Pup Pup, Brian pointed up the track toward the den with his nose when Skelly asked him where Shirley was. She explained to me how she used to tell Brian, '"You go and get Shirley to come down and you watch the babies." Next minute, I see Shirley coming down. I go, "Come on, girl, did Brian tell you to come down?"' After Shirley was killed by other dingoes on the beach in 2015, Brian was both mother and father to their pups.

Skelly said she had seen Brian a few months before when she was sitting on the back veranda at K'gari camp. 'Brian! Brian!' she'd called, and he stopped, came back and sat where he and Shirley used to sit. Another dingo she'd seen a few months previously she thought was Brian, but then decided he looked like Pup Pup because he'd 'got old in the face'.

Like Narawi and Shirley, Sister Girl was a beautiful white dingo, friends with Inky and Winky, Skelly said, but a bit older than them. She was killed by the pack, 'big dogs' near the house at K'gari camp, before the fences were built, and she was buried there. Despite requests, QPWS did not return Inky's and Winky's bodies to be buried at home at K'gari camp.

I didn't want to leave Skelly feeling sad and angry and powerless about Inky and Winky. She was happiest when she related stories like the one about when Sister Girl was trying to get something from underneath a big branch that was lying across the track after a storm. When Skelly approached to get a closer look, she saw Sister Girl was trying to get a pair of sunglasses from under the branch. 'What are you going to do with that?' she asked her. 'Put them on?'

When I listen to the recording of our interview I hear lulls in Skelly's voice. She sounds tired and despondent. Was she ready to stop the interview there? I wonder. Or there? I cringe as hear how I kept going, kept pressing, kept asking questions. Was I looking for what it is, as Skelly says, dingoes know about Aboriginal people? Or is it that I just wanted her to keep talking? I wanted to keep listening to her; to hear how she put things; to be, vicariously, with the dingoes she knew; to laugh with her.

There's another story that Skelly told me twice during our interview, about one day when she and her sister were sitting down talking. Like the day that Inky came to see her and show her his wound, Skelly didn't give me any more markers about when this day occurred. It could have been any day or every day but it wasn't. It was a specific day at a time indeterminate to me. It was the day that Sister Girl looked over at Skelly and her sister from where she was sitting, and joined in their conversation. I think it might be a day that Skelly has relived more than once. Her voice was lively as she mimicked the sounds Sister Girl made and mocked her subsequent conversation with her sister. 'People'll think we're losing it if we say that dingo was talking to us,' she said, but that didn't stop her believing that Sister Girl was talking.

'I'll never forget that day,' she told me. 'She was talking to us.' She sang the sounds Sister Girl made. 'Uh-oo-uh mm-ah … Hmuurr murr.'

I believed her. 'What do you think she was saying?' I asked.

'I don't know,' Skelly said. 'What are youse two talking about?'

Skelly had an appointment at the hairdresser's after our interview. I liked her for that, for the hair dye and the hairdresser's. While she was at the hairdresser's, Rosie and I walked the length of the Urangan Pier. The sun was setting. People were fishing and strolling. Rosie loved K'gari. She told me some of the things she had learnt from the Butchulla – not to spit on the fire and not to whistle at night.

It was dusk when I dropped Skelly and Rosie at Skelly's house but it was easy to follow Skelly's straightforward directions back to the Mantra. When I pulled up in the car park there I found her hair dye on the floor of the passenger seat of the car. So I retraced my route to her place. It was dark now. Unlike the clipped lawns in front of the other houses in the street, tuckeroo trees and callistemons grew in Skelly's front yard. Her two big dogs greeted me. Skelly's house was sparsely furnished, a table, a lounge, a sarong draped over the window. Skelly and Rosie offered me some of the chicken they were eating, they'd had time to buy it just before the corner shop closed, but I wanted to be by myself to let my thoughts run. With the bearings Skelly had given me, I wouldn't get lost in Hervey Bay anymore.

6

Sister Girl

they know more than they admit
Franz Kafka, *Investigations of a Dog*, 1922 [1946]

Perhaps the pack, the usurping juveniles, hated Sister Girl because she was gorgeous, funny and cocky. Do dingoes hate? Perhaps they attacked her because she hung out near the house at K'gari camp and they wanted to be there too. Perhaps they expected her to run away. Did she know, when she encountered them, how far they would go? I can't picture who approached whom. I don't know how the pack came to be on Sister Girl's territory or whether she would have been submissive, crouching and supplicating and licking their mouths. She doesn't strike me as a fighter. Perhaps she whined and urinated. Maybe her display of submission did not appease them. Or maybe she was not submissive at all.

I've seen footage of young dingoes on the island feeding on something on a sand dune when, in a split second, for no apparent reason, one attacked another with convincing vehemence. Their mother intervened to break up the attack.[1] But in this instance Sister Girl's mother did not intervene to save her.

I do not know what injuries killed Sister Girl. In a 2017 analysis of the injuries of fifty-two K'gari dingoes attacked and killed by other

---

1   Behrendorff 2015

dingoes between 2001 and 2016, eighty-eight per cent of them (I calculate that's forty-six dingoes) suffered a subcutaneous haemorrhage in the chest and ribs; sixty-two per cent (or thirty-two dingoes) had been bitten on the neck, and sixty per cent (thirty-one dingoes) on the chest and ribs.[2]

According to this study K'gari dingoes bite and shake other dingoes to kill them. Crushing bites with their powerful jaws and vigorous shaking cause extensive internal haemorrhaging, and more than one-third of the dingo bodies analysed had punctured organs inside their chest cavity.[3] Despite these massive internal injuries, the dead dingoes' external injuries appear minor.

Pictures in Behrendorff et al.'s study of how dingoes kill other dingoes show the body of an eighteen-month-old female dingo before her hide is removed for the necropsy. She lies on her right side, her gigantic ears pointing backwards, her front legs long and back legs longer. She looks deer-like, as if she is leaping. There is not a mark on her coat except, in the close-up, an apparently small puncture wound, caused by her attackers' canine teeth, in the fur on her neck. As her skin is peeled back, the captions list her injuries: subcutaneous haemorrhage (bleeding under the skin) to the left shoulder, ventral chest, thigh and flank area; subcutaneous haemorrhage, intramuscular damage and deep punctures to the dorsal (back) neck region; fractures to the C3 vertebrae and blood in the spinal column from puncture; tears in the stomach lining; and laceration to the liver.[4]

The body of an almost-four-year-old male lies left side down on the sand before he too is peeled and dissected for necropsy. His coat is scruffy, his tail extended, his teeth are bared but not a drop of blood is visible on his coat or on the sand underneath him. There is a puncture wound at the back of his neck. Under his skin are subcutaneous haemorrhages on the right and left sides of the back of his neck, shoulders and back; extensive muscle maceration to his dorsal neck region; and muscle mutilation at the back of his neck at the base of his cranium underneath the apparently small external puncture wound.[5]

---

2   Behrendorff et al. 2017, 5, fig. 1
3   Behrendorff et al. 2017, 4–5
4   Behrendorff et al. 2017, 6, fig. 2A–G

6 Sister Girl

Behrendorff et al. write that when dingoes kill, they kill quickly.[6] The authors do not conjecture on dingoes' motivations for lethal attacks,[7] and, apart from the pup with one toe and the end of its nose missing, they found no evidence that dingoes consumed the carcasses of dead dingoes.[8]

So it is likely Sister Girl suffered puncture wounds on her neck and died of haemorrhaging under the skin in her chest. Her body probably showed no external wounds and left not a drop of blood on the ground she lay on. I don't know what state Sister Girl was in when the pack left her and disappeared into the banksias, sedges, mallee eucalypts and casuarinas of the wallum heathland. Before she died her skin might have felt cold and she might have been thirsty. Her breathing would have been shallow and her legs, 'the pride of dogs',[9] would have been weak. Her heart would have been beating faint and fast. She might have experienced visual disturbances and ringing in her ears.[10] I imagine her blood eddying into her narrow dingo chest, amassing there like a river that cannot reach the sea.

Her thready pulse would have been rapid and scarcely perceptible. If someone had touched it with their finger it would have felt like a fine mobile thread, difficult to feel and easily obliterated with slight pressure.[11] No matter how much her heart wanted to keep going, each contraction would have been getting weaker, pumping less and less blood around her body.

I won't stay with the delta of Sister Girl's blood flowing into her chest. I follow the thread of her pulse to the sea. Her tongue is lolling and her mouth is wide, gaping air. Her lungs are full of air and her body is no longer weak and powerless. Her blood flows to her reliable paws, which touch the ground lightly. She undulates through the air. She reaches the sea and is tossed, dissolved in the cold, salt, crashing white waves. Perhaps the people who love her see her there refracted

---

5   Behrendorff et al. 2017, 6, fig. 2H–M
6   Behrendorff et al. 2017, 4
7   Behrendorff et al. 2017, 8
8   Behrendorff et al. 2017, 4
9   Kafka 1946, 27
10 Farlex 2003–17
11 Merriam-Webster 2017; Amperodirect 2017

in rainbows of spindrift. Sister Girl might have died in December, the month the greatest number of female dingoes are killed by other dingoes.[12] People buried her near the house at K'gari camp.

I don't know what was significant to Sister Girl – her patch at K'gari camp; her relationships with Inky and Winky; the way she died; her interactions with people; or other aspects of her life that I don't know about. I just know she was a curious dingo. It is stretching credulity to think of her as some kind of extroverted, sociable ethnographer. But it's an idea that has lodged in my mind: Sister Girl conducting interspecies research, motivated not just by the need for food or to breed or to defend territory but by her own intriguing, complex reasons. People would think it is preposterous for this fictional Sister Girl to critique 19th-century English ethologist and animal psychologist Conwy Lloyd Morgan's canon, which states:

> 'In no case may we interpret an action as the outcome of the exercise of a higher psychical faculty, if it can be interpreted as the outcome of the exercise of one which stands lower in the psychological scale.'[13]

This view that animal behaviour should always be interpreted in terms of the most simple explanations, rather than those that privilege higher or more complex competencies, became the basis of 20th-century behaviourism. Morgan claimed to eschew anecdote. His aim was to use empirical observation to arrive at objective and non-anthropomorphic findings. In a famous anecdote, he watched how his fox terrier Tony learnt to open the latch to the garden gate and let himself out. Morgan claimed that though Tony's escape may have looked intelligent, he only learnt how to lift the latch by accident. This trick, as Morgan described it, became habitual by repeated association of the chance act in the same situation. According to Morgan, Tony displayed no insight when he opened the latch; it was just trial and error, or conditional, learning.

As far as I know Sister Girl never met a domestic dog. I can imagine her watching, fascinated, her tongue darting out to lick her

---

12   Behrendorff et al. 2017, 7
13   Editors of the *Encyclopaedia Britannica* 2020 (C.L. Morgan quoted)

lips, as she watched Tony opening Morgan's garden gate. Maybe she would have pounced on him while he was attending to the latch. Maybe she would have copied him. Maybe she would have chased him through it. More likely she would have worried away at the latch herself, or found her own way under, over or through Morgan's garden gate according to where she wanted to be.

I can picture her listening, incredulous, head cocked to one side, gaze intense, as Morgan explains Tony's gate-opening ability as behaviour without insight, a chance accident. Perhaps this human need to demonstrate that animals cannot have 'competencies that are similar to [humans'] own'[14] puzzled her. Perhaps this idea is so preposterous she could not even sense the insult in the science, though it is likely she had a sense of what is fair and what is unfair because humans have done studies that show that dogs and wolves are aware of and do not like inequity.[15] Human researchers have hypothesised that canids' dislike of unfairness is closely linked to the cooperative capacities that enable dingoes to hunt, raise pups and defend territory as members of a social group.[16] Focusing on cooperation rather than competition has started to change the ways people think about how humans and canids evolved together in relationships that are mutually beneficial to both species.[17]

I bet Sister Girl appreciated how individual, complex and unprogrammatic her relationships with different humans were. She perceived and reacted to nuances of human expression. She participated in the dances and conversations, the exchanges, the bodily and aural feedback loop of communication with humans. Whatever Sister Girl made of the actions of the people she knew, maybe she would have been too generous to explain human behaviour in the basest possible terms. When she tried to get the sunglasses from under the fallen branch she was, like humans, attracted to novelty. She joined in a conversation with Skelly and her sister. She participated in the comedy of social life and the absurd. Maybe she let humans surprise her and give her pleasure the way she surprised them and made them laugh. Perhaps she believed

14   Despret 2016, loc. 436
15   Essler et al. 2017
16   Essler et al. 2017
17   Pierotti and Fogg 2017

that humans are capable of understanding more about dingoes. Perhaps when Skelly and her sister gave Sister Girl space to converse, when they were willing to listen to her talk, they were all on the edge of discovering something big.

7

Brothers

He couldn't dance his brother out of him, not fully.
Patrick White, *The Solid Mandala*, 1966 [1995]

The day after I interviewed Skelly I arrived early at the office of the Air Fraser Terminal for the 7 a.m. flight to K'gari. Young European tourists arrived too: women in bikini tops and thongs and smooth-faced young men wearing shorts. My feet were already hot in runners. One girl – Danish or Swiss maybe – had crutches and one long leg in a plaster cast. Undaunted by how hard it would be to get around on the sand, I understood from her conversations with others waiting in the terminal that she was returning to the island for a second time.

None of the tourists took a copy of the eight-page A4 *Dingoes of Fraser Island (K'gari): safety and information guide*[1] from the counter. I browsed and kept an eye out for QPWS ranger Dan Novak, who had told me he would be on this flight. Page seven of the dingo education guide described the consequences of habituating dingoes:

> In 2010, a photographer was fined $40,000 and given a nine-month suspended jail sentence for a series of offences related to feeding and attracting dingoes on Fraser Island. The dingoes fed by the

---

1   QPWS 2015

113

photographer, in this case over a period of time, had lost their natural fear of people. They became so aggressive toward other visitors that one dingo savaged a child and, as a pack they cornered fully grown adults – a frightening experience. Although other avenues of management were attempted, these animals continued their aggressive behaviour and had to be humanely destroyed to protect other visitors.

Interesting, I thought, in a dispiriting way, that so much attention was still being given to QPWS's conflict with Jennifer Parkhurst. But not surprising, in the context, to read this justification for killing the Hook Point pack.

I don't know whether dismay is the most productive mood for a researcher but I felt dismay about the seemingly intractable positions of people who knew much more about dingoes than I did. And some trepidation. I wanted to get beyond the confrontational camps, and to do so I needed rangers' perspectives. Perhaps during these interviews the best thing I could do with my sympathy for Parkhurst and my sadness about the killing of dingoes would be to keep them under my cap. I did not want my interviews with rangers to be adversarial, as so much of the conversation around the management of dingoes on K'gari is.

'It is very dangerous to attract dingoes,' I read. 'They are unpredictable and capable of killing people. Don't be fooled into thinking they will react like a pet dog.' An ad for guide dogs played on the TV on the wall behind me. 'I trust them with my children,' a voice on the television said.

As I followed the shorts and long white socks of our impossibly young pilot across the tarmac to board, a man in the QPWS rangers' uniform – dark boots with socks, pale khaki shorts and collared shirt – appeared. I had been reassured by how open and responsive Dan Novak had been in the emails we had exchanged to organise this interview. He was not a big person. He gave me an impression of lightness. But also, among the overseas travellers' bare legs and swimming costumes, we were both people with jobs to do.

Two small planes flew to the island that morning. We climbed under the wings into the cabin of our six-seater Cessna. The woman with the crutches sat at the front beside the pilot. My backpack and

I crammed in behind them. The sheepskin covers on their seats took up valuable room. Novak sat diagonally behind me. He was reading a book, re-reading, he told me, Bryce Courtenay's *The Power of One*. About thirty-five QPWS rangers worked on the island, he said, at three stations – Eurong, Dundabara and a third that I didn't catch.

Below us the Great Sandy Strait shimmered. The water was patterned with tiny dark ripples and the grey-blue shadows of the clouds looked fluffy. On the horizon it appeared as if K'gari were borne aloft by a soft mist of cloud. I doubt we were as high as 2000 feet over the meandering rivers and mangroves of the island's west coast, low enough, maybe, to be able to see dingoes if they happened to come out from under the cover of foliage. We flew over lakes and wooded country at the centre of the island, and then over naked sandblows close to the east coast, before swinging out over the shining path the sun made on the ocean and circling back to the eastern beach from the south. Our plane made a flimsy, toylike shadow over the creek and on the sand near Eurong as we descended. Through the front windscreen the beach came closer and closer, rushing toward us. Commonplace and routine for the pilot, I told myself, but exhilarating and very strange to speed along the beach as a runway. It felt as though we were bouncing over the tyre tracks of four-wheel drives, which, according to traffic regulations, must give way to aircraft when they are landing and taking off.

As Novak had promised, a QPWS ute met us to take us to the ranger station, which was perhaps only a couple of kilometres away from the airstrip but a hot slow journey on foot. I did not look back to see how the young woman with crutches managed on the sand, or where the beautiful tourists dispersed to.

Behind the dune that fronted the beach, among casuarinas and melaleucas that grew out of the sand, the QPWS offices occupied a modest, single-storey suburban-style house made out of hardwood and weatherboards. We entered from a wooden-floored veranda and inside were different rooms accommodating the work stations, computers and filing cabinets of two or three rangers. There were seed pods, snakeskins, animal bones and skulls, dingo-tracking collars and other curiosities on shelves. A parcel of square ear tags – replacements for the round ear tags QPWS were currently using – sat on a long table in the largest room, a kitchen, which had a fridge and sink. Novak told me that only a very small proportion of dingoes were tagged: of possibly 200 dingoes on the island, only fourteen had been tagged between January and November 2015.

We sat at Novak's work station in the office he shared with ranger Ben Steep. Novak introduced us and later, when Steep was about to leave the office to return to Rainbow Beach, I asked how he was travelling. 'On foot?'

'No, in a vehicle,' he replied.

'Not jogging?' I joked, thinking about how fit a person would have to be to run that distance on sand.

'No, no running here,' he said, deadpan, referring to the possibility of inciting a dingo attack.

During our conversation Novak offered me a copy of the *Dingoes of Fraser Island (K'gari): safety and information guide* and I told him I had picked one up at the Air Fraser counter. With it in my bulky backpack was the canid ethology book *How Dogs Work* by Coppinger and Feinstein. Interpretations of dingo behaviour lay behind some of the deep-seated disagreements about the Fraser Island Dingo Conservation and Risk Management Strategy. Talking with QPWS rangers would, I hoped, help me to learn more about dingo behaviour, and how to interpret it. I also needed to understand the steps QPWS staff took before they made the often highly criticised decision to kill a dingo. I doubted they enjoyed killing dingoes. I wanted to know how such harrowing work affected them.

Novak grew up in Childers, a sugar town established in the late 1800s about sixty kilometres inland from Hervey Bay. After his science degree he applied for jobs with QPWS but was told he needed more experience. So he did a twelve-month advanced conservation traineeship in mid Queensland; spent four years managing land, erosion, plants and animals (including feral animals), and doing fire control on defence force land; and worked on geographical information systems for the controversial fox-eradication program in Tasmania. After another unsuccessful application to Parks, he worked for four years as a local council compliance officer in animal management where he investigated dog attacks on people and animals, and managed stray animals and breaches of local council laws.

In one instance, he told me, the police called him to a house they couldn't enter because the dogs were so aggressive. He didn't have words to describe how he had sat beside the deceased occupant of the house to gain the trust of a dog who had been eating the body of its owner. Strange and sad, he said, that someone could be left unnoticed for so long. I didn't expect this story to have a good outcome, but the dog's owner had made arrangements for it and a relative, who knew what it had done – no reflection on the dog, Novak said – took it.

The story reminded me of a scene in Patrick White's novel *The Solid Mandala* in which dogs eat the body of one of the characters. The dogs belong to Arthur Brown and with him they are playful and affectionate. They:

> used to rush out on Arthur, their supple, flashing bodies, their strong white teeth revealed in pleasure, and he would go quite passive, though wobbly, allowing them to lick his hands ... Then he would potter round a bit, talking, or grunting, to his dogs. Mostly in little pleased noises or phrases of gasps.[2]

But to Arthur's twin brother, Waldo, they are threatening:

> Waldo never talked to a dog, and on his arriving home, they would prowl round him, lifting their pads as though they were sprung, and whining through their sharp-looking noses.[3]

In a climactic scene near the end of the novel a neighbour, Mrs Poulter, sees through a dirty window Scruffy sitting on the bed where Waldo's body is '[t]orn by the throat', 'pulling at that other part of Waldo Brown'.[4] The other dog Runt crouches on the floor '[s]wallowing down'.[5] It is a gruesome image but when I read the book my moral sense, my sense of justice and, I think, the plot's sense of justice were not outraged by the dogs' act. The emotional logic of the denouement is easy for

---

2   White 1995, 285–6
3   White 1995, 286
4   White 1995, 302, 303
5   White 1995, 302

people to understand: Waldo Brown, a character who is not kind, died and was eaten by dogs; his twin brother, Arthur Brown, a character who is kind, survived. It does not turn out so well for the dogs. They are shot through the window by the local police sergeant.

Not all of Novak's dog stories ended well. He didn't appreciate being regarded as the baddie and he was angry about being abused given it wasn't his fault that no one came to claim a dog from the pound after two weeks, and he had to take the animal to the vet, sit it on his lap and pat it as it was given a lethal injection.

'My job on K'gari,' he said, 'is probably one of the most heavily scrutinised animal management positions in Australia … but coming to work in such a magical place and trying to conserve and look after dingoes, as an animal and a species, is wonderful. We do have to euthanise an animal occasionally. Two this year, which is still a terrible experience for us, but on the whole I think Parks are doing a really good job here.'[6]

When he started to talk about the Eurong pack, and show me photos of them on his computer, I was enthralled.

'This particular animal,' he said, showing a picture of an eight-month-old male dingo taken in March 2015, 'was ripping into people's tents. Not being aggressive, not snarling. A guy left two loaves of bread in his tent and he broke into it and stole the two loaves and for the next four nights in a row ripped into four other tents looking for more bread.'

Novak named the Eurong dingoes by the colours on their tags: 08Red and his brother PurpleGreenYellow (PuGY) who were born in July 2014, their mother 08Purple and her mate, their father, PinkYellowPink (PiYPi).

I asked him whether rangers ever called dingoes nicknames to simplify their jobs.

'No,' he answered, 'we try not to. It's bad luck for them, I think. The animals that get named are very, very habituated. In the past they have been named by locals and people feeding them. They just become a problem animal and they're the ones we've had to euthanise. So we try

---

6   Novak 2015. Subsequent direct quotes from Novak are from the same interview.

to just keep it to their tag rather than treating them like a pet. They're not a pet.'

Across the room Ben Steep talked on the radio while Novak described the keyhole surgery a dingo like 08Red might perform on a tent. 'It starts off with people leaving food in tents and he can smell it very well. They will rip into the corner where the food is sometimes, rip a little hole, grab the food and off they go – not even get into the tent.'

08Red's father PiYPi – Novak said the words 'pink yellow pink' – was an example of how a dingo should be: a magnificent hunter, a fantastic, wonderful animal, with respect for people. He didn't do anything wrong, he moved away from people. He managed to navigate his heavily touristed territory and had picked up only a couple of Code C reports for loitering.

But when young dingoes like 08Red procured food from humans they were, in the eyes of QPWS, establishing bad habits that did not give them the skills they needed to hunt for themselves. 'Every time he goes tent ripping,' Novak said, 'he's not out with his father. He's satisfied. He's going to lie down in the dunes and he's not going to go with Dad and find a bandicoot.'

Unlike PiYPi, 08Purple, the mother of the Eurong pack, was, as Novak put it, 'very, very habituated', which means, according to the Fraser Island Dingo Conservation and Risk Management Strategy, that she had lost her 'natural fear of humans'.[7] Born in 2013, 08Purple gave birth to her first litter, which included 08Red and PuGY, at just a year old, which was young because female dingoes do not usually breed until they are two years old. Novak described her way of acquiring food from humans: from the dunes, 08Purple would watch a fisherman thirty metres away down on the beach as he put bait on his line. When he turned his back to cast off she would run down the beach, knock over the bait bucket, grab the packet of pilchards and race back up to the dune, all while the fisherman had no idea.

She was teaching her 2015 pups, who were about four months old at that stage, to do the same. 'All six of them up there watching her.'

---

7   Ecosure 2013, 7

Despite the bad mothering 08Red received, Novak didn't regard him as a nuisance animal. 'He will just opportunistically target food that people leave at their camps and in tents. That's about all he does.

'But this animal's brother had four Code Es, where they actually make contact with a person or hurt a person or ambush people. They're quite scary. Two of his Code Es were where he nipped two different people and he was removed on August 16.'

'I read about him,' I said.

This dingo, called PuGY by QPWS, was the one I had nicknamed Bold. I didn't tell Novak I had met him. My one-minute encounter was paltry compared to his contact with the Eurong pack. I was here to learn from the rangers' experience. I had read many incident reports about Bold/PuGY as well as the press release QPWS issued after he had been killed. Sitting beside Novak at his work station I could not recall the details of the incidents. I had had trouble absorbing and putting them into chronological and spatial order. Perhaps these difficulties were related to my desire to meet QPWS rangers without presuppositions; I wanted to hear their perspectives as they presented them to me.

So I read aloud from Novak's screen a Code E interaction report about a series of incidents that had taken place one day in early July on the beach at Eurong. PuGY had run at a group of people; lunged at a woman 'attempting to bite'; circled a group of three people and a ranger's vehicle; and propped and lunged. He was not dissuaded by a ranger with a long-handled shovel, or put off for long by the pumice the ranger threw at him, which he grabbed with his mouth before he ran off toward two adults extracting pipis on the beach at the water's edge. The ranger followed PuGY in the vehicle to where the dingo was trying to separate the pipi collectors and was pushing a person into the water. There must have been liquid detergent in the QPWS vehicle, which was squirted at him. A ranger stalled him while the group of three and the two pipi collectors made their way to the fence that surrounds Eurong. After a couple of minutes PuGY ran toward the five people only to be headed off by the arrival of another QPWS vehicle. Nevertheless, he again displayed dominant behaviour and persisted in circling around the vehicle. The five people had reached the inside of the fence before he ran north toward the main entrance to Eurong.

'Dominant/submissive testing' and 'dominant toward humans' are classified as Code D or threatening behaviour on the dingo interaction reports. Displays such as a dingo dropping on its paws and crouching, then springing up; lowering itself to the ground and propping; bouncing around on its front paws and lunging; running toward people with its tail curled over its back; running back and forth in front of a person; propping and jumping from side to side; jumping and yapping; and lunging are regarded as dominance testing. Dingoes are regarded as showing 'dominant toward humans' behaviour when they are not easily deterred or scared away. PuGY exhibited all these behaviours numerous times.

While humans often see such actions as aggressive, they are also part of dingo play. So the interpretations of the dingo's motivation are important – especially if humans are seeking ways to de-escalate threatening situations. In my interviews with rangers I wanted to learn how humans can distinguish dingo play from dingo aggression, especially as dingoes' lives depend on humans' abilities to discern how to gauge the situation and act accordingly.

In ethological terms, dominance occurs when one individual supplants another in the competition for food or breeding opportunities.[8] Perhaps it is this form of dominance Novak referred to when he made the case that when a human gives food to an adult dingo, the dingo sees that as a form of submissive behaviour. Other ideas about dominance are drawn from observations of canids' social behaviour, in which an animal's posture indicates whether, as part of the group, it has dominant or superior status; inferior or subordinate status; or whether its status is somewhere in between.[9] The position of a dingo's ears, whether its mouth is opened or closed, whether its tail is held high or low and how it moves, and how much eye contact it makes are all indications of its ranking in the group.

In dingo groups a low-status individual shows submission by cringing, avoiding eye contact, wagging its tail, licking the muzzle of the higher ranking dingo, or, if threatened, rolling onto its back in an act of trust. Instead of avoiding humans and disappearing if they came into his territory, PuGY ran straight toward people with his tail curled

---

8   Appleby 2015, 144
9   Smith 2015, 38–9

over his back. Around humans he opened his jowls wide, crouched, pranced around and vocalised. These actions indicated that he did not understand his position in the human–dingo hierarchy.

From ripping tents without people in them to ripping tents with people in them to scenes like the one on the beach, PuGY's juvenile delinquency had progressed to sub-adult criminality. The interactions had been increasing in frequency from once or twice a week to once a day to several in one day.

'That dingo,' Novak said, 'was responsible for nineteen out of twenty interactions in the Eurong area for that month, and fifty per cent of all interactions throughout the entire island for that quarter. Out of the 200 or so dingoes on the island, one dingo was doing half the interactions. Another thirty or forty per cent of the interactions were in his territory but not identified. So probably close to ninety per cent of interactions on the whole island …'

It was PuGY's overweening persistence that earned him a Code E for the events I had read about on the beach at Eurong.

'What was he doing there?' I asked.

'He was pushing the limits of his relationship with humans.'

'What would have been his objective?'

'He was testing to see what we are. Whether we're a threat. Whether we're a prey item. Whether he has any concerns. Whether he can target us to get food … '

Naturally, dingoes don't regard humans as prey, Novak said, but he told me he had seen dingoes prey upon cattle three or four times his weight, and capable of defending themselves. 'So why not a person? Dingoes cause horrific injuries to people,' he said. 'Intelligence and wariness are humans' only advantages.'

'When he bit,' I asked, 'were they severe bites?'

'No. We're quite lucky that in all of those cases there were people that could intervene.'

He told me about another Code E incident in which two girls, aged around eleven and seven, stood back to back on the beach at Eurong – as advised by Parks – while PuGY circled them four times, eventually at a distance of less than one metre. When one of the adult onlookers saw him nip at one of the girl's pants twice he ran at him with a poly-pipe rod holder. Deterred, PuGY headed off south along the beach.

'That circling, is that testing?' I wanted to locate the line that distinguished curiosity, possibly playfulness, from aggression.

'Oh, listen, that's just predatory behaviour, looking for an opportunity, a weakness.' When multiple dingoes circled people, he explained, one would race in from the back and grab a person on the back of the legs.

'That circling the car, separating people, is it dominance testing?' I was asking Novak to do what I could not – explain PuGY's state of mind. Why did he keep propping and jumping from side to side and circling?

'Occasionally it's dominance testing. It's also exploratory testing as well, finding out how far they can push their limits,' he explained. 'It's learning about us, and learning techniques, and probing, which is what dominance testing is. It's educating themselves about what they can get away with.'

In Novak's view, humans were culpable: PuGY had been fed from when he was a small pup. He took responsibility: Parks had failed to stop people corrupting dingoes and turning them into something that they're not.

PuGY's first recorded nip occurred on 23 July 2015. By then he was known as a problem dingo and he had been fitted with a radio-tracking collar so he could be identified and monitored. He surprised a woman by sniffing the back of her leg. She panicked and ran. He grabbed. 'It mostly likely would not have happened if she'd listened to our dingo safety messaging,' Novak said. 'The worst thing you can do is run.'

Running is dangerous because it activates canids' inherent predatory behaviours. According to canid ethologists Coppinger and Feinstein, each time dingoes hunt, capture, kill and eat prey, and many times when they are not successful in the hunt or they are just fooling around, they deploy a specific set of discrete motor patterns in an integrated sequence. They *orient*, that is they locate prey and turn toward it; they stand still and *eye* and smell their prey; they lower their body and move slowly toward their prey in a *stalk*; they run full-speed forward toward their prey in the *chase*; they *grab-bite* to disable prey; they *kill-bite*, often an artery, to kill prey; they use their teeth to *dissect* tissue and expose internal organs; and they *consume* their prey. Components of this sequence, for example, the *stalk*, the *chase*, the *grab-bite*, normally occur in a particular order so that grab-bite usually occurs after chase.[10] As the ethologists put it, 'it appears that chase is a releaser for grab-bite'.[11] So if a person runs, the dingo's subsequent grab-bite is part of an intrinsic motor pattern. If the chase is interrupted or delayed, the grab-bite is less likely to ensue. If a person stands still and does not run, a dingo is less likely to grab-bite; the sequence of predatory behaviours is less likely to be activated.

The collar enabled QPWS to map PuGY's movements and, Novak told me, did not affect his ability to travel. He showed me a photograph of lines on a satellite image of the island that tracked his expeditions up and down the eastern beach north and south of Eurong for many kilometres and inland. He journeyed nineteen kilometres from Dilli Village to Central Station in an afternoon. A tangled ball of lines near Eurong revealed where he spent time at the family den with his brother, his mother, his father and the 2015 pups.

Novak collected evidence to prove that the collar did not affect his behaviour or other dingoes' attitudes toward him. In the first of a series of photographs of PuGY and his father that Novak took the day after he was collared, PuGY is standing side-on to the camera, looking small and puppy-like. He cringes. His tail is down but not fully between his legs because it irrepressibly loops up like a roller coaster and kinks to the right at its tip. His ears are back, which contributes to the impression that he is small; his face is turned away from the camera toward PiYPi. PuGY's tongue protrudes horizontally from his muzzle, ready to supplicate. He is dripping with submission. I can't see his eyes but everything about his posture indicates that they would be forlorn and imploring.

'Submitting means safety,' Novak said.

PiYPi, front on to the camera, looks like a demon. His jaws are open and his teeth are bared, his canines and lower teeth clearly visible. They form a formidable frill of fangs around the dark red cavity of his mouth where his tongue lies. His eyes are fluorescent, pale green pinpoints. They are reflecting light back to its source from his tapeta lucida,[12] special surfaces between his optic nerves and his retinas that

---

10   Coppinger and Feinstein 2015, 64–97
11   Coppinger and Feinstein 2015, 82
12   Latin for bright tapestries

enable him to see better in low light. In the photo these retroreflectors give him a supernatural aura, not that he needs one because he is larger than life already, drawn up tall from his pale paws on the sand in the dune grass. His chest is narrow, his rear haunches are muscly. His ears are erect, neither forward nor back, but clear and imposing against the backdrop of casuarinas behind him. He occupies all the space he can. His magnificent tail is high, with the pale underfur visible, and flopping to his right.

In the next photo PuGY's nose is close to PiYPi's. His head is down, swinging sideways to approach his father. His ears are still back and his tail curls horizontally to his right. PiYPi has closed his mouth and lowered his eyes and his muzzle. His ears are less upright; his tail has swished a little to his left. He does not look angry. It seems to me he might be looking tenderly at his son.

In the last picture the two dingoes stand with their rumps close to each other, their bodies making a thirty-degree angle from the point where they connect. PiYPi's body faces the camera. His head is turned to his left, his jowls apart as if he is smiling, relaxed. He looks younger, less serious. PuGY's body is at an angle to the camera but his face is turned toward it. He is licking his lips almost sheepishly, his ears askew,

his dark eyes nearly closed. Phew, his face could be saying. They have one tail visible.

Footage Novak took two days after PuGY's collar was fitted shows how well ranger and renegade knew each other. 'He would know me better than I know him,' Novak said. He told me he thought a dingo would have to be pretty close to hear a human's heartbeat, but PuGY could scent the same detail as humans see.

Novak was sitting down eating his lunch when PuGY approached him. He stomped his foot and the dingo retreated for a moment, then opened his jowls and waggled his head, vocalising. Still interested in Novak's food as Novak stood up and said, 'No!', PuGY thrust his nose forward, sniffing, then circled around before leaving to approach some fishermen. After they packed up he scavenged a few little bits of bait left on the beach before he went to the next group.

Inevitably, during an interview, I do not think quickly enough. It did not occur to me to ask some questions I realised were important later. The pictures of PuGY and PiYPi demonstrate that PuGY functioned normally as a member of dingo society. He understood his father's communication and responded appropriately. If PuGY had made it through another year, to the age of two, he would have matured, developed a respect for people, Novak told me. Why, I wondered later,

did humans' attempts to communicate with him, to deter him, fail so spectacularly? PuGY was a big, strong-looking dingo, 18.6 kilograms and only a year old but Novak didn't think he had the potential to be the leader of a pack, to breed. He doubted he could hunt for himself, all he was doing was targeting campsites. He was, according to Novak, a 'very local animal'.

I was vitally interested in 08Red, the quiet brother, the proof that dingoes are as individual as humans, and that being fed by humans – which results in what the Fraser Island Dingo Conservation and Risk Management Strategy defines as habituation – could not account for everything. PuGY and his brother 08Red grew up together around Eurong. They emerged from the same den and they ranged over the same camping zones, the same stretches of the beach. They were good mates. Smaller than his brother and not the dominant one in their relationship, 08Red was, perhaps, an unlikely survivor. On one plane negative characteristics, what he did not do, enabled him to survive: he did not approach people as relentlessly as PuGY; he was not as persistent; he did not nip. On another plane positive attributes might ensure his survival: Novak observed that 08Red was more independent, explored more of his territory than his brother.

I do not know how much 08Red knew of what happened to his brother, what he made of PuGY's disappearance. But the question of what he did after PuGY was killed is not beyond the realm of human understanding. Novak told me that around late October, when the 2015 pups were about three months old and becoming more active, his parents, PiYPi and 08Purple, booted 08Red out of their territory. One of the rangers had seen him the day before our interview, hanging out, not misbehaving, on the beach just south of Indian Head, more than sixty kilometres north of Eurong.

Novak predicted some hard times ahead for him. 'Over the next couple of years, twelve months,' he explained, 'he's going to try and establish himself in a pack up there, be accepted into another pack, and survive.' With the mayhem and furore around PuGY over, my thoughts went with 08Red, slipping away to make his life in a strange new brotherless world.

At the end of *The Solid Mandala*, Arthur Brown talks to his neighbour Mrs Poulter about his brother, Waldo, who has died. 'I don't think … I could live without my brother. He was more than half of me,' he says.[13] Mrs Poulter assures Arthur that Waldo was no more than a small quarter. Soon after, she pleads with the policeman who is taking Arthur away to be kind to him. 'Kindness is something he understands … This is a good man, Sergeant. You know it in your heart.' Sergeant Foyle is sympathetic to Arthur Brown but he is also a functionary of a justice system that has its own logic. The circumstances are beyond individuals, and what they see and what they know in their hearts. 'It's not a matter of hearts, Mrs Poulter,' Sergeant Foyle replies, 'The issue is something to be decided by better heads than mine.'

---

13   White 1995, 311

8

What they're capable of

if there were no enclosures the very first dingo would savage me
Vladimir Nabokov, 'A Guide to Berlin', 1925 [1996]

While Dan Novak and I pored over pictures of the Eurong pack, a ranger came in with two take-away coffees. She gave one to Novak and he asked her what it was. 'Lucky it's not poison,' she joked. I was impressed by her gall and didn't realise straight away that this was Linda Behrendorff. I had wanted to interview Behrendorff since I first started to learn about the dingoes on K'gari. Even among people critical of Queensland Parks and Wildlife Service, she had a reputation as one of the better rangers. Her help was acknowledged in two of the most informative scientific articles I had read about K'gari's dingoes.[1]

On Novak's screen was footage of PuGY lying on his side after Parks staff had trapped him for some routine procedure. He was around nine months old and he looked very young. His muzzle and his legs were no longer taped up, he was free to go, but he didn't move for what seemed to be a long time.

'As it should be,' said Behrendorff. The rangers explained he was checking everything was okay before he moved.[2]

I thought he was petrified.

---

1   Allen et al. 2015; Appleby et al. 2013

They explained how a dingo submits when it feels threatened. Firm pressure – Behrendorff put her hand on my arm to demonstrate – actually relaxes a young dingo.

I asked all the people I spoke with in the course of my research whether they had ever been afraid of a dingo. Novak told me he had not been afraid but he had exercised caution when he came across four dingoes feeding on a dolphin carcass. He was checking on a trail camera on the dolphin when four hungry dingoes lifted their heads up, looked at him from the top of a dune and formed a semi-circle around him. He backed away.

'I'm surprised how often people ask me, "Are dingoes really dangerous?"' he said. On Novak's computer screen were images of a human with dingo-inflicted wounds. 'They just don't understand they're very, very capable of causing injury.'

While he took a phone call, Linda Behrendorff took over interpreting the photographs of the German backpacker who had been attacked near K'gari Educational and Culture Centre on 28 July 2012, 'before the fence'. She spoke quickly and softly. She had a matter-of-fact intonation that reminded me of the way a police officer might talk on television about a tragic house fire or brutal family violence: concerned; professional; informing the public, for whom they work and to whom they belong; but their account circumscribed by a complex set of unseen factors.

'From what I can gather and from interviewing him, he was looking for a toilet block. He was intoxicated, two in the morning, and he wandered up the Woralie Road the wrong way,' she said. 'There were some habituated dingoes in the area that we had concerns with. They had no fear of humans and they were causing tent rippings and nipping, and they were being protected by some elements of the traditional owners up there that considered them their camp dogs. So the contributing factor in this case is the habituation. They don't have a natural fear of humans. They've lost that.'

She told me that when the backpacker couldn't find the toilet block he sat down on the road and snoozed. He woke up, as Behrendorff put

---

2   Behrendorff 2015. Unless specified otherwise, subsequent direct quotes from Behrendorff are from this interview.

it, 'to his face being bitten and everything being bitten'. She explained how, first, dingoes test: 'They'll come up and mouth you. Nyeah, mouth, mouth.' She said 'mouth' with the vowels in middle long with a falling intonation and the final consonants short, like a dingo vocalising. 'Then it starts pulling hair. And then it might start nipping. For whatever reason it's then turned into full-on nipping.'

Behrendorff told me the backpacker was so drunk he wouldn't have felt the dingoes until the attack started to become very serious. When he played dead, the dingoes kept coming. So he started fighting back. He punched the dingoes but they continued to bite him, teeth coming at him from everywhere.

'The only part of his body that wasn't bitten or damaged,' she said, 'was next to his groin and stomach area because he had that covered up most of the time. Lucky they didn't get into his femoral artery. These areas here,' she indicated his lower legs on the screen, 'you can actually see his tendons and his fat through his skin. They'll do this to calves and that.'

'How many was it, do you know?' I asked.

'It was pitch black. He'd lost his thong, he'd lost his torch. He'd tried to run up the hill. He tried to climb a tree while they were attacking. He tried to get away. He kept falling down. Eventually he's got himself up a tree and then stayed there. He doesn't know how long. And then he got back down for however long, he couldn't remember. He sat there and then he's fallen asleep again.'

Behrendorff described many of the backpacker's actions in the present perfect tense – he's climbed up a tree; he's fallen asleep again – a grammatical construction favoured in police accounts, too, possibly because it links past actions to present repercussions.

Her description of how the backpacker woke up at dawn with 'a dingo face right in his face' is funny, but would have been scary for the backpacker, because, according to Behrendorff, he thought the dingo would attack again. The man stayed as still as possible while the sun came up. The dingo stayed there, too, looking at him, tongue out, with 'a kind of smile', as he told Behrendorf. Another dingo stayed in the distance.

The backpacker got up and started walking up the track. When he realised he couldn't hear the sea anymore he turned around and went back the other way. One dingo followed him closely. Another followed at a distance. Without harming him they accompanied him back to K'gari camp.

After interviewing the injured backpacker, Behrendorff went to the place he had described. She found a flattened area, the site of the initial attack. She found his torch and his thong up the dune. She found bits of his jumper. He told her the dingoes grabbed his jumper and tried to pull him down so he took it off and let them have it. It was all roughed up on the ground when Behrendorff found it. Everything she found verified the backpacker's account.

'The only prints I could find,' she said, 'were about two prints because we do print measurements. Probably wouldn't stand up in a court of law, but two to three animals attacked him and two to three animals took him carefully back to home. And there's only two sets of prints. That's how Jekyll and Hyde it can be.'

I have listened to the recording of this section of our interview over and over again, pausing and replaying it as I transcribed and summarised, as I wrote and read and rewrote. Behrendorff admitted she didn't have photographic evidence of how many dingoes and which dingoes attacked the backpacker. Her circuitous answer to my question was based on the paw prints QPWS found: she believed the dingoes who escorted the injured man back to camp the next morning were the same dingoes who had attacked him.

Behrendorff visited Justin the backpacker in hospital in Hervey Bay. He was just about to go in for surgery. Three deep wounds on his head, more on his calf. 'He was lucky he didn't lose his eye,' she said. 'He was lucky to be alive. If they'd wanted to kill him he would be dead.'

She scrolled through more photographs on the computer. 'This one here sticks in my mind. Anzac Day 2011. That's a three-year-old girl.'

I do not have a good memory of the pictures QPWS had kept of the puncture wounds on the little girl's legs. I didn't want to look at them.

The girl's parents were among a group of sixty people waiting at Hook Point, the southern tip of the island, for the barge to the mainland. They didn't keep a 'full eye' on her and she got up and walked over the dune. 'Those particular dingoes were waiting,' Behrendorff said. 'For want of a better word, I guess, the audacity of a supposedly wild dingo to come in and strike that child. The dingo grabbed her and pulled her down. Another male dingo came in behind her. Just missed her femoral artery. The barge skipper could see the dingo going for her and yelled through the barge loudspeaker.'

Forensically, Behrendorff listed the contributing factors: it was breeding time; the child shouldn't have been so far away from the group. But the dingo was not skinny, not even lean, so, apparently, not hungry. Behrendorff was appalled that the dingoes did not leave the site after the attack. People rang to say they were still milling around. Behrendorff instructed the observers to take photographs and to keep the animals in their sights. She sent two staff down. 'We're straight onto it.'

Two dingoes, a male and a female, both almost three years old, were killed by QPWS at Hook Point on 26 April according to Parks' *Humane destruction database*.[3]

Years before, the female was known to Parks. She was one of the 2008 Hook Point pups, the pack Jennifer Parkhurst had visited. In May 2009, when four of her littermates were killed by QPWS, 'They let her go because she hadn't bitten enough. She didn't have enough form, wasn't enough of a risk at the time to be humanely destroyed with her siblings,' Behrendorff explained. 'And she went to ground. They hadn't seen her for a while. And she's come back at this age. It's the only one that I've ever seen it happen with. It doesn't normally happen. At two years of age they're gone. They pull their heads in and don't start this again.'

The attack on the three-year-old girl by an adult dingo blew Behrendorff away. 'She was a mature female from a pack that was highly habituated from a very young age. Whether the attack was because of the full-on habituation she had from the den due to the wildlife photographer, whether that's got any bearing on it, but it definitely is an anomaly. She was habituated from the den. Parks don't go into the den site. We don't have anything to do with the pups. We might put a trail camera up or something but we don't have anything to do with the pups. We're not there. We're not in their face. These animals were daily visited.'

On Novak's screen was a picture of the female dingo taken shortly after the attack. 'She's actually got bits of blood around her muzzle,' Behrendorff pointed out.

It is strange to look at a photograph of a condemned animal. Looking at this photograph in this context would, I thought,

---

3   Allen et al. 2015, supplementary material

unavoidably join my perspective with that of the dingo's prosecutors who believed that this dingo was to blame and that violence against this dingo was justified. Pictures of animals used and killed by humans so often take away animals' dignity. I did not want to rob her of her dignity. I felt that looking at a picture of her so soon before she was killed might be a macabre sort of voyeurism and a continuation of the human violence she had known since a young age. Part of me wanted her to have peace and some sort of privacy. Part of me wanted to study her very closely, as if I could learn what was going through her mind.

'Zero fear of humans there at all,' said Novak.

'Didn't care,' said Behrendorff.

'She was euthanased,' I said. 'Do you think if a dingo did that they'd do it again?'

'Who wants to take that chance?' said Behrendorff.

No stranger to controversies about dingo management on the island, she was careful to say only what she knew. But I was constantly wondering what was going on for the dingoes. A week before the attack, on 19 April, an almost ten-month-old male dingo was killed by QPWS at the Coolooloi camping area.[4] I don't know if this young dingo was the offspring of the two who mauled the toddler. I don't know if his death had any bearing on their behaviour.

Later, after talking with Novak and Behrendorff at the QPWS offices, and after going for a drive along the beach with Novak, I went with Behrendorff to replace a traffic counter at the Champagne Pools car park. As she drove along the eastern beach with the calm skill of long experience, I asked her what would happen if QPWS didn't take out the dingoes who had form in the interaction reports, the ones who had bitten or mauled.

'This is what risk mitigation is about.' She explained Parks and Wildlife's job was to conserve the population, and to conserve the habitat so all species on the island could survive. They had a duty of care to people as well as dingoes. The QPWS ute rose and fell rhythmically over the sand moguls made by the northerly wind. 'Is it the sacrifice of the one or two,' she said, 'that actually saves the rest of the population?

---

4   Allen et al., supplementary material

Nobody wants to see the removal of another thirty-two dingoes if somebody gets killed.'

She was referring to the aftermath of the fatal mauling of Clinton Gage on 30 April 2001. Later that day a QPWS ranger killed the dingoes suspected of attacking Clinton, a male and a female estimated to be twenty-two months old, when the pair arrived back at the Waddy Point camping area.[5] Behrendorff said that they were 'very, very habituated, camp dogs'. A post mortem revealed that the male's stomach contained no items of human origin, though dental impressions of his teeth showed patterns that corresponded with Clinton's injuries. The female dingo's stomach contained suspicious material but its origin was uncertain.[6]

The order to kill came from the top, from then Queensland Premier Peter Beattie. 'No politician,' Behrendorff said, 'nobody in authority is going to stand back when a child has been killed. They got people in to do it,' she said. 'It's not something anybody wants to do.'

They were killed in ones and twos and threes: on 2 May a ten-month-old female at Waddy Point; on 3 May two females, one ten months old and one almost three years old, at Central Station; two males at Dilli Village; one female at the Wanggoolba barge landing; a male and a female at Lake Boomanjin; and a male at Eurong – all almost three years old. On 4 May an almost three-year-old male was killed at the Wanggoolba barge landing, another at Dilli Village, a male and female at Hook Point, a male on the Woralie track and another at Wyuna Creek. It continued. On 5 May nine dingoes were killed: one at Central Station, one at Lake McKenzie, one at Kingfisher Bay, one at Eurong, one more at Waddy Point, one at Moon Point and three at Indian Head. From 30 April to 13 May 2001, thirty-one dingoes were killed.[7]

In 2001 Behrendorff was a public-contact ranger. She lived at Central Station, a former logging camp between Eurong and Lake McKenzie, one of the island's most visited lakes. Dingoes lived under the old hardwood forestry barracks where Behrendorff lived. Every morning at around five a female dingo howled under the brushbox floorboards of her barracks room until Behrendorff got up and

---

5   Allen et al., supplementary material
6   Appleby 2015, 136
7   Allen et al. 2015, supplementary material

acknowledged her. When Behrendorff told her to go away she went off to howl outside the tents in the adjoining camping area. 'You knew their routine. You knew what they did. They were part of your life,' she said.

The stable pack of dingoes at Central Station knew the humans' routine, too. Behrendorff didn't think they relied solely on human food, they still hunted, but it looked to her as though they regarded campers' food as easy pickings. 'They'd be back at lunchtime, they'd be back at dinner time. They'd go wandering and do what they do, but unsecured food – Christmas cake, bread – was fair prey.' I extrapolated that they might regard campers' food in their territory as their own resource. Behrendorff explained how the dingoes killed in the April–May 2001 cull were the ones who lived in tourist areas and had been exhibiting behaviour similar to the behaviour of the dingoes who had attacked Clinton Gage. 'They were all dingoes who had the potential to do what those other dingoes did,' she said.

Many, including Aboriginal people, members of the Fraser Island Defenders Organisation and other conservationists, were appalled by the killings. There was a public outcry. 'I know there is criticism of our decision to cull – I would simply point out that a young boy has died,' said Premier Beattie at the time. 'We take no pleasure from this but we have a duty of care to residents and visitors. This is the right thing to do.'[8] As Queensland Environment Minister Dean Wells put it during a television interview, 'When a little boy died the whole world changed.'[9]

After Clinton Gage's death, QPWS started in earnest the system of incident or interaction reporting. In 2002 dingoes started to be ear tagged. Prohibitions on contact between people and dingoes, and feeding dingoes, were more strictly enforced. Some people, who thought the dingoes attacked Clinton because they were hungry, advocated setting up feeding stations, a policy rejected by QPWS because it would alter dingo population dynamics and artificially increase the carrying capacity or the numbers of dingoes who could live on the island. Behrendorff raised questions about where supplementary feeding would lead. How would it affect dynamics within dingo packs and between packs? And when would feeding end? How much food

8   Alfredson 2001
9   Channel 5, 2001

would be the limit? Behrendorff argued that dingo fatalities are part of life on K'gari. Dingoes die from many causes including snakebite, attacks from other dingoes, starvation. 'Not every dingo's meant to live,' she said. 'We've just got to suck that up. It's not nice. That's what's been happening for thousands of years.'

'Why do people get so emotional about dingoes?' I asked Behrendorff.

'It's been an emotive subject since [the case of] who took Azaria,' she said. 'It divided a nation. It put an innocent woman in jail.'

She told me how she had felt concern, not fear, one night when she was walking on her own in the dark from the old forestry barracks at Central Station up to her accommodation in the duplex at the top of the hill. She walked this path every night after dinner with a work colleague and had known the alpha female in that area, who often followed her for a stroll, for quite a while. But this particular night the dingo was 'in a mood' – agitated and twitchy. She came up behind her, followed her closely. Behrendorff kept talking: 'I know you're there. Cut it out. Don't even think about that.' She had no torch to see her way, no food that would have attracted the dingo and no idea what had precipitated the dingo's animosity. If she hadn't kept talking she thought the dingo probably would have come closer and bitten her on the back of the leg.

'You can learn to read them.' She spoke about developing rapport, assessing the vibe. Sometimes, she said, 'You know it's not a good time to approach, not a good time to be there. You don't know what has happened before. There are so many factors to take into consideration when you're dealing with these interactions. Maybe they've come from a fight.'

In the QPWS offices at Eurong that morning I had shown Behrendorff and Novak a picture of a wolf 'gaping' – opening its mouth wide and showing its teeth with its ears back. Did a dingo's wide-open mouth indicate aggression or an intention to attack, or could it be, as the ethologists wrote, 'possible that the animal could be using the gape as part of play'?[10] Pictures in the interaction reports showed PuGY with his mouth wide open. Behrendorff did not want to comment on a photograph – the animal in the picture could have been yawning; a video was more informative. Dingoes did come in and open their

---

10   Coppinger and Feinstein 2015, 64, fig. 13

mouths very wide and 'carry on', without snapping, she told me. 'I do see that behaviour affiliated with what people call dancing or that prancing, paw to paw, jumping around.

'I can't tell you what it means,' she continued. 'Whether it's a test or whether it's waiting for a response from you. I can't read their mind. I'm probably not qualified to give behaviourist definitions. It would only be my opinion. It could be, "I'm showing you my teeth." It's not a snarl.'

But if their lips were pursed back and their teeth were bared, she told me she would read that as aggression, that the dingo didn't want her in their area, and she would retreat.

'Retreat with confidence,' Novak interjected, 'don't allow the animal to think it has dominance.'

'What if a person just stands there?' I asked.

'Depends on the dingo,' Behrendorff said. They were as variable and as various as people. The story was in the whole animal and its movement she explained. 'The tail tells a story. The eyes tell a story. If the tail's up, especially if hackles are up, that's something to be worried about. Hackles are down, tail's sort of swirling around, to me that's just curious: "I'm just having a go here. I'm bored and I'm going to try you out." Their behaviour is a function of their hunting instinct,' she said, 'but it doesn't mean they are going to eat you. However, when people back themselves into the water, it's exactly how they teach their young how to herd up a kangaroo. "Back it into the water, eventually it'll get so tired, we'll just wait here."'

Novak said that he was confident that a person could easily change their interaction with a dingo by standing up straight and taking a step toward the dingo and not freaking out.

Behrendorff advised standing one's ground, being authoritative. She, personally, would look it in the eye and change her voice – 'Don't you even think about it!' – so the dingo heard it as a growl. She had never had to grab-punch or eye-gouge a dingo to protect herself, but defending oneself vigorously is the recommended action if attacked.

'If you're attacked,' Novak said, 'you're defending your life and you need to defend it aggressively.'

Sensitive to the seemingly endless feedback loop flowing through dingo–human interactions, Behrendorff said, 'You're dealing with an animal that is intelligent and you're dealing with a human that's got their own intelligence and perception of things. So when you get an interaction you're dealing with the reaction of the person to the initial action of the dingo that will cause whatever outcome it's going to be. On the other hand, you're dealing with the reaction of the dingo to the action of the person.'

Later I ponder possible reasons humans feel so ambivalent about dingoes. Dingoes react to minute and unconscious signals humans emit; they reflect and challenge humans; humans are unprepared for dingoes' scrutiny, attracted to and afraid of that level of involvement, unused to such limbic negotiations. Dingoes force humans to confront a novel form of responsibility based on not being wholly in control. And sometimes humans are oblivious to the field of negotiation.

9

Traces

'Know about dingoes, do you dear?'
'Enough to know I don't know enough.'
John Heffernan, *Chips*, 2001

After we'd talked in the office, Dan Novak drove me to One Tree, a camping zone just north of Eurong, where Parks allowed a commercial tour operator to leave eight empty tents set up. The food-preparation area was fenced. 'We lock the backpackers in there,' Novak said. 'Dingoes can have the other 180,000 hectares.' Commercial tour operators were permitted to use One Tree if tour groups prepared and ate food in the fenced area, if guides reported all dingo observations and interactions, and if every guest was briefed on dingo and human safety. The two camping zones closest to Eurong, One Tree to the north and Wongai to the south, had been closed to the general public since August and would stay closed until March so that the 2015 Eurong pups could move around their territory with less likelihood of coming across campers and their food. Novak wanted them to learn to target bandicoots and goannas and stuff washed up on the beach rather than Oreos and Cheezels.

Our QPWS ute bumped over the sand undulations and I was lifted out of my seat. Novak apologised when my head hit the roof. We passed a stationary four-wheel drive, its occupants attending to a flat tyre. Novak had had more flat tyres on the island than he would have expected.

I asked him what would be the ideal way for dingoes and people to co-exist here.

An ideal interaction, he told me, was a neutral one, not one in which dingoes were rewarded with food from people. He was frustrated that people would tolerate a dingo like PuGY hanging around their campsite for days and only report him when he stole some cheese. 'That person, by enjoying that wildlife experience, has contributed to that dingo becoming habituated.'

I sat in the ute beside Novak, trying to process this information and apply it to my experiences on K'gari. Was it wrong for people to be curious about dingoes, to observe dingoes? I couldn't work out how to *be* around dingoes. We were in their territory. Didn't they have a right to be here and if they wanted to observe people, didn't they have a right to do that too? Wouldn't that involve people looking at them, enjoying looking at them?

'Is it better for people to observe dingoes in a zoo?' I asked.

The best scenario, he said, was for people to stay twenty metres away from dingoes. If they were on the dunes, wind down the car window and watch them from a distance. Use the zoom lens on a camera. Stay out of their faces.

My meeting with Bold, or PuGY as Novak called him, was playing through my mind. It seemed impossible to me to stay in the car all the time. How could people walk, fish, set up a camp?

'But what if a dingo approaches a person?' I asked. 'Is scaring it off the best thing a person can do?' Not for the first time I thought about my daughter, who had just turned nine, and how terrified she would be if a dingo approached her. I knew how my heart would pump and my stomach would drop and my spine would freeze.

'If you're too scared to scare it off, the dingo knows that, and you should stay in your car or stay with someone that is confident to behave correctly."

Walk in a group, Novak said. He recommended carrying a stick: he found dingoes tended to maintain a distance just further than the length of the stick. He told me people with children could stay in one of the fenced camping areas at Central Station, Dundubara, Waddy Point and Lake Boomanjin.

The dune and its grey-green foliage passed in a blur on our left. At a seemingly random spot along the seemingly endless beach, Novak turned off up a track. We had arrived at another place. Cornwells. It was not a camping area that an urban person might imagine but a series of sandy patches behind the dune that fronted the beach. I didn't take photos. I wish my memory of the place was better. The vehicle track seemed indistinguishable from the rest of the ground, the light diffused through the spindly foliage of the casuarinas. It was a place PuGY and 08Red knew, the place they were tagged. It was a place the Eurong pack continued to frequent because Behrendorff had radioed us to say she had just seen some dingoes there.

Novak pointed out the electric fence around the barbecue area. There was a trailer where another commercial tourism operator kept their food locked up. He found the commercial tour operators, CTOs in QPWS parlance, more diligent in securing food than the free and independent travellers, called FITs by the rangers. People often underestimated what dingoes can do, he said. He described how they could unclip the latches on a Waeco fridge, lift up the lid and enjoy the contents; they could climb up on the roof of a car if people put their rubbish there. Last week dingoes had broken into an independent camp site and stolen a big bag of roast meat and dragged it up to the commercial tour operator site where they sat in the shade and ate it.

We saw paw prints but no dingoes.

We continued driving, talking about how to secure rubbish (put it in the fenced skips QPWS provides) and how not to appear submissive to a dingo (don't turn your back), when Novak spotted something on the dune. He stopped and we got out of the ute. A pile of rubbish buried in a shallow hole had been dug up. Torn plastic, jagged tins, synthetic-coloured packaging, oozing secretions, human 'hygiene' mess: sloppy but not yet putrid. Of immense interest to a canid. Paw prints all around. The Eurong pack, Novak said, PiYPi and 08Purple. Their 2015 pups learning that humans were a source of food. All this litter – plastic wrap, meat bags, canola-oil containers, Coke cans – smelt like people and their campsites.

'Why would you do this?' Novak's exasperation was clear. It was more effort to dig an ineffectual hole than to take the garbage to one of the QPWS skips. We put on rubber gloves and picked it up.

* * *

I didn't time my drive with Novak. It may have been late morning when we returned to the QPWS offices. Behrendorff needed to replace a traffic counter at Champagne Pools car park, an hour's drive north of Eurong, and I accompanied her. We passed the four-wheel drive with the flat front tyre and she pointed out the entrance to the walking track to Lake Wabby, south of Cornwells. I thought of the vehicle as a space where we might be able to talk more openly than in the office. It was private but we were still out there, in all the magnificence of the eastern beach.

As Behrendorff drove she told me she spent her high-school years in Maryborough. Her father's family came from there; when he was twelve years old her step-grandfather had worked on one of the boats that brought logs from the island. She studied zoology and protected-area management remotely. Before she started volunteering for QPWS on K'gari in 1999 she worked as a zookeeper with Australian native animals. She found working in zoos a little bit heartbreaking, she said – the lack of money, the need to use animals for public interest. Tour-guiding, showing people animals in the wild, using her personal experiences to elaborate on how they feel, smell, interact and work was a challenge she relished more. She started working full time on a casual basis for Parks in 2000, and lived on the island for ten years until 2009. Now she flew in and flew out each day.

She had thought about what it meant to be a wild animal. When I asked her what a wild animal is, she explained: 'A wild animal has potentially the freedom to do what it wants or needs to do in its wild capacity when it wants to do it.' Though some wild animals, like koalas, were constricted by urban boundaries, she elaborated, captive animals have quite strict boundaries and sometimes their behaviours are altered.

Behrendorff had shown me a photograph of a five-week-old dingo pup walking down a vehicle track, seemingly oblivious to the large four-wheel drives that, she told me, nearly ran him over on a blind corner near Eurong. He was dark, as the K'gari pups are born, and much bigger and more developed than a domestic dog of that age – with long, strong-looking legs; an oversized head, muzzle and ears; and a skinny little upturned whip of a tail – but still clearly a young pup even though his small torso was starting to show the lean shape of an adult dingo. Where did he fit in the wild/captive categories, I asked Behrendorff. He was neither wild, living with limited human interference, nor captive, restricted by pens or cages. Behrendorff described him as free ranging. He was habituated, she said, displaying no fear of humans, because his family had been associating with humans, not necessarily for food, for generations. The mother of that pack, her mother's mother, her mother's mother's mother and father, the whole family line, had lived at the interface of campers, the resort and the residential valley of Eurong.

In Behrendorff's view the Eurong pack didn't necessarily want or need to be around humans. 'It's just, "You're there, we don't have an issue with you, you're just part of our territory, part of our area." I would assume, I would hypothesise,' she added quickly. 'I can't speak for them.'

As for myself, since I had started extrapolating from my dog's perceptions to thinking about dingoes, I had tried and failed many times to resist the compulsion to attribute feelings, thoughts and utterances to dingoes.

'I know you can't predict,' I said, 'but what's the likely life course for that pup?'

'In the time I've been here, I can only go by what I've observed. The course for a habituated pup from a habituated family, it will either survive and once it gets to two years of age go off and do its own thing or it could be influenced by human factors, whether it be feeding or familiarisation, and possibly end up on that path down to humane destruction,' Behrendorff said. 'It could be hit by a car. It could be taken out by another dingo.'

Behrendorff said that one or two of the females would go on to have their own litter and continue the cycle. She told me she had noticed that the females tended to stay in their home area. In contrast to earlier research on a captive dingo pack in which the dominant female killed the pups of any other female who whelped,[1] she had observed members of the Eurong pack – sisters and an old girl called Granny – cooperate to

---

1   Corbett 1988; 1995, 81–2

raise three separate litters at the same time. It was in the past, before the fence went up around Eurong in 2008, when people and dingoes still mingled. She had no evidence of infanticide, though she did not know what happened in the dens.

'If QPWS's objective could be achieved and people stopped intentionally and unintentionally feeding dingoes and interacting with them,' I asked, 'how many generations would it take for the Eurong pack to become unhabituated?'

Behrendorff told me she had noticed changes. Before the fence, she said, there were some dingoes called 'the dump dogs' who took their growing, more mobile young to the dump out the back of Eurong when they emerged from the natal den. The pups were left there, on old mattresses and in discarded dongas, while the adults went out to forage. For the last two years the dump had not been used. The last generation of pre-fence dingoes had disappeared or died.

When we were somewhere between Happy Valley and the coloured sands near Dundubara, Behrendorff told me about being nipped by a dingo.

'Did it hurt?' I asked.

'Yes,' she said and laughed. 'Of course it hurt. Their teeth are sharp but they're blunt, so even if they don't pierce the skin they'll cause a crushing sensation. It affects your nerves underneath. The bite mark stays. Put it on the record,' she added, 'there were no repercussions for that animal!'

In 2009 she and animal behaviourist Rob Appleby were putting a collar on a dingo as part of a study on aversive conditioning.[2] Behrendorff described how they had a dingo in a cage, sedated. They were using a leather welding glove as a kind of pillow so he didn't rub his newly tagged ear up against the cage. Behrendorff was holding the other glove when the caged dingo's brother came up and nicked (her word) the glove .

'He took off, did a big loop, dropped the glove and came back out of the bush,' Behrendorff said. 'I thought, *I'll go and pick it up.*'

She described how she took her eyes off the dingo as she extended her leg to step over a log. As she brought her other leg up, he came in

---

2   Appleby 2015; Smith and Appleby 2018

and nipped her calf. She picked up a stick but he was too quick and flung back.

'Then he looked at me as if to say, "What are you going to do now?"'

I asked Behrendorff whether she thought the dingo's behaviour was an instantaneous, senseless, random action – a brainfart – or whether he was trying to communicate something.

'You can read anything into it. We had his brother in a cage coming out of sedation. Anything could be going through his head. Could have been standing up for his sibling. Who knows? It's not up to me to go into that. It's a whole arena of what-ifs and hypotheses.'

'Why would he steal the glove?' I asked.

'Because that's what they do!'

At the entrance to the Champagne Pools car park Behrendorff found the spot in the sand by the tyre-grooved track where the traffic counter was buried and dug it up. I'm not exactly sure what she needed to do – take a reading, replace batteries, replace the traffic counter itself – but whatever it was, she did it competently. We stood for a while on one of the platforms on the boardwalk overlooking the pools, which were named for the effervescence created by waves breaking on the rocks as they filled them up with water, but the tide was low. My phone recorder was switched off. I was concentrating not on the immensity of water, light and air around us, the salty rocks and the wind-beaten foliage, but on what Behrendorff was saying. She was talking about Jennifer Parkhurst, 'the photographer', as people who worked on the island called her. 'I asked her and asked her to stop feeding dingoes,' she said.

Parkhurst and Behrendorff must have been on reasonable terms once. Parkhurst had told me that Behrendorff's interaction reports were like 'chalk and cheese' compared to those of some of the other rangers. I interpreted that to mean that she thought they were more considered, more accurate, fairer, more honest. Caution, an awareness of the limits of her knowledge, shaped Behrendorff's approach. 'You can't just go in with one view, you've got to find out the whole scenario. You ask this person and that witness and you can get a very different story. "It nipped me on the ankle and tried to bite my ankle off" can turn into "I accidentally kicked it in the mouth as I was running away from it" and vice versa. The dingo can change from male to female, from a blue tag to a red tag. It's a big job trying to get the truth and the facts.'

Behrendorff described how she compiled interaction reports from what she witnessed and from her interpretations of the data people gave her. It was important to allow informants to describe what happened. She didn't ask leading questions but she would return to a question, ask it three times in a different way at different times in the course of an interview. She saw the interaction reports as a way of managing and mitigating risk. 'You're not trying to get an animal into trouble. You're not trying to negate an actual high-risk situation by playing it down.' If QPWS staff knew a dingo had nipped, they could do something about it, she said. They could close a camping zone, keep people away from a particular area, monitor the dingo.

The category of severity of the report, Code C – nuisance; Code D – threatening; or Code E – high risk, was very important. QPWS asked long-term dingo experts Lee Allen, a zoologist who works on pest animals for Biosecurity Queensland, and Laurie Corbett, an ecologist who specialises in vertebrate predators and the management of feral animals, to assist with the definitions of Code B, C, D and E interactions when they developed the interaction reporting system in 2001–2002 after Clinton Gage was killed. In 1998 Corbett had predicted the 'distinct possibility of a human death from dingo attack' on the island.[3] Behrendorff had recently asked Lee Allen for more clarification around the Code D and E incidents because, for example, there were only minute differences between a Code D lunging (not attempting to nip or bite) and a Code E lunging (attempting to nip or bite), which, in the panic of the moment and without being cognisant of a dingo's intentions, would be very hard for a person to assess.

For Behrendorff the interaction reports were useful because they were more systematic than memory or diary entries. Rangers knew where the hotspots were, knew the risk in certain zones and could allocate more resources in those places to dingo safety education. But she had argued against having a simple quantifiable system, a form of mandatory sentencing, in which, for example, two Code Es would result in a specific and consistent action on the part of Parks because situations and dingoes were so variable. A biting Code E was different from a less aggressive Code E. Ripping or entering a tent without people

---

3   Corbett 1998, 12

in it were Code Cs; ripping a tent with a person in it and entering a tent with a person in it were Code D incidents. But sometimes, Behrendorff said, a dingo might simply be attracted to the smell of a person's fishy hands on a tent guide rope; obsessed by scent, they might enter a tent without even knowing a person was in there.

She pointed out that what one person might perceive as dingo aggression, another might regard as play. She told me she had long discussions with Rob Appleby about these issues and that, according to Appleby, in all the many hours he had spent filming dingoes on the island, they had not been aggressive toward him. 'But,' Behrendorff said, 'just because it didn't bite you doesn't mean it isn't being assertive.'

Habituation is another topic of debate in relation to the management of dingoes on K'gari. Behrendorff explained how, from what she had observed and from the data QPWS had, dingoes familiar with humans, dingoes who had lost their fear of humans, dingoes defined by QPWS as habituated, were the ones involved in high-risk incidents. 'Habituation doesn't always lead to assertiveness or aggression,' she said, 'but a high percentage – not all – of high-risk interactions are from dingoes known to QPWS who come from packs that have grown up around resorts or homesteads or areas frequented by humans. Habituation is the common denominator.'

Behrendorff told me that Appleby did not agree that habituation led to aggression. 'He doesn't see where that can be defined,' she said. I assume that she was referring to the lack of peer-reviewed, published scientific evidence showing a causal link between habituation of dingoes on K'gari with aggression. She wanted to move on from opinion and assumptions and she appreciated Appleby's critique. She, like me, was doing postgraduate study – her thesis was on dingo diet and ecology. The two research articles she had under review when we met in 2015 are both now published. One, about dingo longevity, includes intriguing photographs of a female K'gari dingo who had lived to the age of thirteen and borne numerous litters.[4] The other is about the variety of dingo diet on the island.[5] She told me that she wanted to write a paper on the dingoes who had not been killed by QPWS, 'the

---

4   Behrendorff and Allen 2016
5   Behrendorff and Allen 2016

animals we've saved' as she put it, the ones who, as juveniles, were the subject of interaction reports but who had subsequently disappeared from the human radar. She wanted to see the valuable data compiled by QPWS over the past fifteen years put to good use. She wanted the public to get another factual view of what went on so they could make informed decisions about the way dingoes are managed on the island.

We had time to visit the carcass of a melon-headed whale (*Peponocephala electra*) that had been washed up near Eurong about six weeks before, in mid October. Near a creek on another stretch of seemingly nondescript beach, Behrendorff stopped the ute and we walked up the dune. 'Look for the bollard,' she said. Novak and Behrendorff had told me about the whale. Behrendorff was writing about marine strandings in her paper on dingo diet. Between 150 and 200 kilograms, it provided, as Behrendorff put it, a 'massive big mountain of food' for wildlife. She was exasperated that the Eurong dingoes had been munching on the whale for weeks and the pups still looked so skinny. 'They've got ribs sticking out and hips sticking out,' she said. 'They look like they haven't eaten for a week to some people'. The siblings still fought, she told me, dominance tested, over the best bits.

The whale was smaller than I imagined, perhaps a few metres long, just ribs now with skin attached to its tail. Dingoes had tugged its head away from its body to a place out of range of the trail camera, which was mounted to a casuarina tree. 'Buggers,' said Behrendorff, who had been going through hours and hours of trail-camera footage for her research. 'What's really frustrating is when they've swung it around, the tail's over there just out of shot and you can see the whole skeleton moving and you know there's a dingo there but you can't write that down because there's not one in the shot.'

The whale's many teeth were still in its mouth. Its stomach had been full of decomposed fishing net, Behrendorff said. Later she showed me pictures and footage of 'who's who' around the whale. A goanna and a sea eagle negotiated where they were going to feed. The sea eagle flew around and came back to one end of the whale, away from the goanna. The dingoes left it until it was putrid. After the tight skin on its skull had exploded and its guts were spilling out, one of the pups ate the maggots from its mouth.

Parks brought marine strandings over the first dune, if they could, wherever they washed up so they didn't become traffic hazards, and so their stink didn't waft over the camping zones in the south-easterlies and wildlife could feed in peace. Impossible for me to imagine the whale, also called the many-toothed blackfish, a highly sociable member of the dolphin family, swimming and hunting in deep tropical waters with hundreds of others and feeding on squid. Now it was a stranding; its details and the fishing net in its stomach had been documented and sent to a state-wide database. Now it was, as Behrendorff described it, an 'opportunistic feeding station'.

Back at the QPWS office Novak, Behrendorff and I sat in the kitchen and talked. Both Novak and Behrendorff thought QPWS were doing the best they could to keep people safe and to preserve the lives of K'gari's dingoes. 'The criteria for high-risk dingoes are a lot more stringent on the side of the animal now,' Behrendorff said. 'So you're giving the animal the benefit of the doubt without playing Russian roulette. There are systems in place. You've really got to be able to identify it and make sure there's no opinions there … When the final hammer falls, so to speak, there is a huge process all the way up to the executive director [of QPWS] and in some cases the minister [of the environment]. You've got all the history of that animal, everything it's done.'

Novak thought the decision to remove PuGY was the best one. 'I was afraid that the next interaction we had with that dingo was going to be a serious mauling, especially in the vicinity of Eurong where we've got small children around. It was an 18.6 kilogram apex predator. It knew exactly what to do. It had already killed a number of dingoes in the area – it was definitely involved in the death of some dingoes from another pack. They are very efficient at killing.

'We had some animals that came over here from other packs,' Novak continued, 'and they were killed here on the beach in front of Eurong – just as it's possible for PuGY to get killed if we moved him up to another part of the island.' Extrapolating from Novak's comments and the six dingoes listed in QPWS's deceased records from 1 January to 25 August 2015,[6] I guess that PuGY may have been involved in the deaths of a nine-month-old female dingo near Eurong in April and a ten-month-old

---

6   Allen et al. 2015, supplementary material

male south of Eurong in May. Both Novak and Behrendorff told me that even though dingoes kill other dingoes regularly, carcasses have shown no evidence that dingoes eat other dingoes on K'gari.

Some of the most indelible things I heard that day were not recorded: walking back to the QPWS ute out on the beach I suggested to Novak that the management strategy had failed PuGY. He answered, 'We're not looking after one dingo. We're looking after a whole complex community.'

Sitting on an eroded part of the dune on the eastern beach, waiting for the afternoon flight back to the mainland, Behrendorff recounted the horror of Clinton Gage's death. 'He was as tall as me,' she repeated. The breeze cooled us and the shadows of the casuarina fronds danced on the sand as she she talked softly about the dingo killing that followed. 'I've let it down,' she said, speaking of any dingo killed for high-risk behaviour. 'I've never put an animal down that I haven't felt empathy for and a sense of loss.' The trauma – for Clinton Gage and his family, for Behrendorff, for K'gari's dingoes and those who care about them – was as palpable as the sand around our feet.

10
Meeting at Pialba

Dingo and moon, those two made a culture. Moon said: 'I'll stop three, four days dead, and I'll [be] coming back again!'

Walaku [Dingo] said: 'I'll got to die forever. I'll lose my self, bone and dust ...' [...]

Ngumpin [humans] been follow that dog way. Where we going? We'll die. Walaku, he'll die too, dog, well he got a two life ... When he die he got another life.

Daly Pulkara in Deborah Bird Rose, *Dingo Makes Us Human*, 2000

I woke up the next morning regretting the photographs I didn't take – the melon-headed whale, the pile of rubbish, the rangers. I had been intent on listening, asking questions, concentrating on words. I recorded the interviews on my phone. I didn't take another camera. I was immersed in the moving world – the northerly breeze, the fluid sand, the fronds of the casuarinas drooping like hands waiting to be kissed; I couldn't freeze these sensations in photographs. My senses did not have the cohesion of novelist Paul Auster's canine character Mr Bones in *Timbuktu*, who was able to distil a place to its essence:

Chicago was a bus splashing through a puddle ... *Tampa* was a wall of light shimmering up from the asphalt ... *Tucson* was a hot wind blowing off the desert, bearing with it the scent of juniper leaves and sagebrush, the sudden, unearthly plenitude of the vacant air.[1]

K'gari was competing voices, my trying to remember what was important, the absence of Bold.

I had timed this trip to coincide with a meeting organised by QPWS and SFID to view footage of Bold. It was a bright morning and there were, maybe, about thirty people gathered in the large room at the community centre at Pialba where the *Dingo – Friend or Foe?* forum had taken place in May. Most of the attendees must have been SFID supporters I surmised when audience members voiced their questions and comments because they showed a definite antipathy toward QPWS. I felt a bit sorry for the three Parks personnel, who seemed to have been roped in at the last minute. The senior Parks managers who organised the meeting weren't there because they had gone to Brisbane to collect an award instead. Senior Conservation Officer Naomi Stapleton had been camping on the island with conservation volunteers till yesterday and had only been told that she would be attending the meeting with SFID that morning.

Jim Peckham,[2] regional director for the Sunshine and Fraser Coast, told the gathering that QPWS only euthanased a dingo that was presenting an unreasonable risk to visitors to the island. Parks assessed the level of risk and were showing a higher risk tolerance than they had in the past. 'We've moved the balance as hard as we can to one side,' he said.

SFID's priority was to find out how many dingoes were on the island, whether the population was viable and whether there was enough food available. Dingo researcher Adam O'Neill was worried about the high mortality rates from 2012 to 2014. He asked whether dingo aggression might be related to stress – could a scat study investigate levels of stress among K'gari's dingoes?

Peckham cited the numbers that have been used since Laurie Corbett made his first rough population estimate in the 1990s –

1 Auster 1999, 33
2 A pseudonym has been used here and subsequently.

between 100 and 200 depending on the season.[3] The Department of Environment and Heritage Protection had provided $50,000 for research; QPWS supported this research and would supply data if requested, he said. He explained that more studies weren't done because of the limited resources. It wasn't cost-effective for rangers to become scientists, he said. Senior Ranger Ben Steep concurred. It was better to separate management from scientists.

In response to the hostile tone of some of the questioners, Stapleton pointed out that all rangers were trained in natural-resource management. 'We do this job because we are conservation-minded,' she said. At the time I thought she was saying we are all on the same side; but now I think about it perhaps she meant Parks were more capable of looking after dingoes.

Steep didn't want to hypothesise; Parks needed researchers to make accurate estimates based on the data available. A recent vehicle strike of a dingo, Peckham assured attendees, was being investigated as a very serious matter, the same level of investigation as pollution of the groundwater aquifer.

Karin Kilpatrick from SFID asked why QPWS didn't keep a puppy register. There was a death register. We need something to measure mortality against, she said. People weren't seeing dingoes like they used to. 'Where are the dingoes?' she wanted to know.

Steep answered that rangers tried to minimise contact. They only monitored dingoes in areas where they were habituated. He reckoned that QPWS had done a good job of reducing the number of habituated animals. 'Look at the dingo tracks,' he said, to indicate there were still plenty of dingoes.

There was discussion about the wildlife-care facility that SFID wanted to see on the island. Sea World, an amusement park on the Gold Coast, was willing to train personnel and SFID had received many donations to set up the facility. It seemed to me to be a debate between ecology-minded people who thought nature should take its course and animal-welfare-focused people who were concerned that animals needed prompt treatment and pain management.

---

3   Corbett 1998, 7

Finally someone played the film I had come to see, footage that Neil Cambourn, the executive director of QPWS regional operations east, had taken of Bold on 1 July acting, according to Cambourn, in an exceptionally erratic way.[4]

It started in the middle: PuGY/Bold close to the person filming him, head low, ears forward, eyes on the camera. His body and face were in shadow, white-tipped tail curled lopsidedly over his rump, leaning to his right. He paced in front of the camera, hardly leaving an imprint on the hard sand, which was damp from being below the high-water mark. His shadow stretched out toward the camera, toward the men watching him, toward us viewers, from his big white paws. He walked in a semi-circle in front of the person filming and stopped. He lifted his nose up and down, scenting, took a couple of steps forward and, as the man behind the camera spoke to him, 'Gotta settle down, matey,' his wide pink tongue briefly curled over his top lip.

'He's in good nick,' one of the men said.

'Big dog, eh,' said another.

Bold looked away, lowered his head, and opened his mouth as he raised and lowered his head.

'Don't even think about it,' the man behind the camera said as the dingo paced back in front of him.

Bold lowered his head and vocalised, a gruff deep 'hruh' sound.

'Behave yourself,' the man said.

'Hroo, roouh,' Bold retorted, tossing his head and lying back on his haunches as though settling in for a conversation.

'Behave yourself,' the man repeated. 'Go on.'

Bold sniffed the sand to his right, looked as if he was going to roll, decided not to. Mercurial, he lifted his head, faced the camera, mouth open. His lower carnassial teeth were visible only when I slowed down the footage as I played it on my computer. I paused the next instant: he had closed his mouth and his foreshortened muzzle pointed straight at the camera. His lips looked thick, a little contorted as though he was mid utterance. His eyes were very wide, very dark. Concern? Fear? Aggression? Imploring? Aggrieved and demanding? His address was unreadable.

---

4   A still from this footage appears on the cover of this book.

Always moving. Always his mouth in front of him. He remained lying down, and, as if he couldn't bear his own honesty, he spun his head around to sniff his haunches, as if to defuse the situation. Then he mouthed the sand. He raised his snout to vocalise and his lower carnassial teeth were visible again but he was not facing the men. He was looking away, where something off to his right briefly took his attention.

The men were talking. I couldn't hear what they were saying. Bold's head spun forward, he play bowed as he rose from the sand and sprang to his left, moving like a gusty, fitful wind. He paused as he looked past the men down the beach, turned his head to gaze to the sea, a moment of stillness amplified because his gestures, stance, expression, posture, movements were constantly changing, as though every scent molecule he inhaled, every sound he heard, everything he saw precipitated a different action. Each movement was 'an expression of a thought, a feeling, a plan, an urge'.[5] He licked his lip, then returned for his close-up, vocalised in a low voice, wagged his head and made a drawn-out noise – 'Uroh' – as he semi-circled the men, side-on with his head toward them.

He was a nervous dingo. He was a bunch of nerve endings.

One of the men said, 'Git out!' and he bounded a few metres away, play bowed, and lurched this way and that before the footage blacked out.

This quicksilver performance lasted fifty-four seconds. I was pleased to see him animated, alive, filling up the room in Pialba. I didn't perceive his behaviour as aggression but to understand what he was communicating would be 'like learning how to speak a new language'.[6]

The rangers at the meeting hadn't previewed this footage.

'Don't infer,' Steep said, 'that this was the behaviour that got him euthanased.'

Peckham agreed. 'We do not destroy an animal for that sort of behaviour.'

When someone asked whether this dingo was part of a pack or whether he was a loner – there had been plenty of stories about him – Stapleton said she didn't know much about this dingo.

'He shows no fear of humans,' Steep said. He did know about this dingo: he was part of a pack; he was seen with other dingoes.

---

5   Auster 1999, 37
6   Auster 1999, 37

'That animal's behaviour was just typical juvenile behaviour,' said Ernest Healy, an academic from Victoria who owned dingoes and was a member of the National Dingo Preservation and Recovery Program. 'Dingoes are an ancient semi-domesticate,' he continued. 'They are not a wolf.' There was a serious conceptual flaw in the current management plan, he said, that habituated animals were not natural and wild. The plan, according to Healy, was an attempt to achieve the unachievable, trying to keep a semi-domesticate away from hordes and hordes of people.

Someone asked why there hadn't been problem dingoes years ago, before 1992.

Peckham replied that data was being captured now that wasn't previously captured. The rangers were working with tag-along tour operators; their camps needed to be fenced. Parks issued a cap on commercial tour operators and on campers on the island.

Ecologist Arian Wallach described how curious dingoes, who had never encountered people before, approached her on her field trips in the most remote places in Australia. 'You cannot control the dingoes and you can't be responsible for the people,' she said. People in the room were angry. 'It's a policy decision,' she said, addressing the rangers. 'Your situation is impossible.'

'At the moment,' Peckham said, as though he had many other things to do, 'we are totally captured by managing the interface between dingoes and people.' If they hadn't taken actions to manage the situation in 2001, he explained, QPWS would have been liable.

Stapleton talked about all the messaging, the communication.

'You can't go to the toilet without getting dingo education,' Peckham contributed.

'Normal behaviour,' Healy said, 'is categorised as aberrant, but that is an institutional lie.' He wanted policy honesty: 'How do we manage normal dingo behaviour as a risk to the public?'

There were refreshments after the meeting. I introduced myself to Naomi Stapleton, thanked her for her help when I had requested to interview rangers in May. I figured that I could not hide my presence at this forum from Steep, whom I had met during my interview with Dan Novak at the QPWS offices in Eurong the day before, so I tried to catch his eye and say hello. We stood in the same circle of people but he did not look in my direction.

The community centre at Pialba is a modern, golden brick and white-render building surrounded by neat gardens and smooth roads. The room where we met was cheery and full of light. At times some members of SFID were antagonistic toward the QPWS rangers, which might be fitting because Pialba is, according to some linguists, derived from the Butchulla word 'baiya-ba', which means fighting place.[7]

But in 1937 Miss R. Bromiley wrote to the *Maryborough Chronicle* with a different etymology: her father told her that the original Aboriginal name for Pialba was 'Barilba', which, she said, was confirmed in a pamphlet by Edward Armitage, a 19th-century settler familiar with the Butchulla language. According to Bromiley's father, Barilba became Pialba when an early surveyor could not pronounce the 'r'. She had a hazy idea that the meaning of the more musical Barilba 'had something to do with the sound of the tide flowing in among the rocks'.[8]

Pialba/Barilba's importance was spiritual, according to medical practitioner, anthropologist and collector L.P. Winterbotham who, in the 1950s, compiled some of the customs and beliefs of the Aboriginal people of south-east Queensland. According to Winterbotham, the Butchulla buried their dead on the day they died, lying straight with their feet toward the west and their arms by their sides in the sand. They believed that on the second day the spirit came back. On the third day relatives of the deceased went and camped at a rock called Pialba/Barilba where Birral, their ancestral being, had left the mark of his left foot as he leapt on his way to the sky.

The spirits of the dead followed Birral to the sky country from this rock. From their bough shelters on either side of the rock, two healers (*gundir*) watched to see the spirit 'jump off'. If they saw it, they lit a fire to make smoke so the spirit didn't come back and frighten people. The deceased relatives gashed themselves with eugarie or sharp pipi shells, and threw boomerangs.[9]

Maybe, with springs in his strong legs, Bold jumped off too, to a country where lack of understanding is not so catastrophic.

---

7   Bell and Seed 1994, 165
8   Bromiley 1937
9   Lauer 1977, 13–4 (L.P. Winterbotham quoted)

11
Wongari

They yarded that wild ocean
to be lakes and swamps for the people's fishing, they lay down
around the old men on a cold night
    Les Murray, 'The Sand Dingoes', 1996 [1997]

That afternoon Finn Dwyer[1] had already bought a coffee by the time
I arrived at the outdoor cafe where we'd arranged to meet, at the Eli
Waters shopping centre, on a busy corner in a suburb of Hervey Bay.
Dwyer was a Butchulla man who worked a nine-day fortnight as a
ranger with QPWS on the island. He told me he loved working on the
island, which, he said, 'is a special place for everyone, like a magnet that
draws you in'. When he came home on his five days off, he said, 'You
know you're loved'.

The first thing he wanted to say about dingoes was that they were
very important to Butchulla people and to all people, and that K'gari
was their country as much as everyone else's.

'They were our companions back in the olden days,' he told me.
'Our ancestors, they roamed together. When Europeans came, they
took that companionship away so now they're on their own.'[2] In the

---

1   A pseudonym has been used here and subsequently.
2   Dwyer 2015. Subsequent direct quotes from Dwyer are from this interview.

past, he explained, two types of dingoes existed: camp-dog dingoes or pet dogs, called *wat'dha* in Butchulla, and rogue or wandering dingoes, called *wongari*.[3]

'Wongari,' Dwyer said. The emphasis was on the first syllable, which rhymed with 'song'; the second syllable rhymed with 'car', the third with 'me' and the 'r' sounded like a 'd'.

I tried to imitate.

A lot of Butchulla words, he explained, are about the sounds and movements and gestures of different animals, though he couldn't say what the source of *wongari* and *wat'dha* was. For instance the word for horse is *yaraman*. 'Yaraman, yaraman, yaraman, yaraman,' he said with the rhythm of galloping hoof beats, the 'r' sounding like a 'd'. He wasn't so sure about the pronunciation of *wadja/wat'dha*, the pet canid. 'Wad-jaa,' he said. 'Correct me if I'm wrong.'

I couldn't correct him. I didn't know.

When Dwyer said K'gari I didn't hear the 'K', just the 'g'. The first syllable was long, it rhymed with 'starry'. Other people had corrected my pronunciation of K'gari when I said it in a way that rhymed with 'Harry'. I might have been making the same mistake as 19th-century Europeans who called K'gari Owens, one of the three Butchulla men who had travelled to Victoria to track down bushranger Ned Kelly, '*Garry* Owens'. K'gari Owens (or Werannabe, also known as Barney) has thousands of descendants. Other Butchulla families trace their descent to Harry Aldridge, son of an early settler of Maryborough. In the 1870s and 1880s Aldridge grazed cattle and horses on his lease at Eurong. He married a Butchulla woman called Lappy and they had four children, after he and Lappy's older sister had had a daughter.

'Does K'gari mean paradise?' I asked.

K'gari is the female white spirit, Dwyer explained, who was sent down with Yindingie by Birral, the supreme high spirit, to help create the country. Dwyer described how K'gari was Yindingie's apprentice. She watched him make the bay and Mount Bauple and she fell in love with the place because it was so beautiful. She asked Yindingie for

---

3 *Dictionary of the Gubbi-Gubbi and Butchulla languages* spells the words *wadja* and *wangari* (Bell and Seed 1994, 135, 136). This book follows the spelling used in a 2017 QPWS brochure: *wat'dha* and *wongari* (QPWS 2017, 2).

permission to stay but he said, 'No. You are a white spirit. You can't be part of this.'

K'gari begged and begged.

Eventually Yindingie relented but, he said, 'You can't stay here as a spirit.'

So he formed her into an island and dressed her in clothes, which are the trees, gave her eyes to look back up to the sky country, which are the lakes, and gave her animals and people to look after so she would not be lonely.

'Her voice sends a message through all—' Dwyer broke off and smiled at a woman and a child at the far side of the cafe. 'There's my wife.'

I smiled at them too and watched them walk away.

'K'gari's voice is the running creeks. All the creeks over there are her voice,' he continued.

When I asked whether the white spirit K'gari was related to the white spirit dingo that I'd heard about he said, 'No. The white spirit is like our mother. She's one of the helpers that came down, Yindingie's apprentice. Yindingie is the rainbow serpent, spirit man from the sky, a man of many colours, not just for Aboriginals but for all living things. In my eyes he's like Jesus.'

Country owned the dingoes on K'gari, Dwyer told me, no one else. The island was the dingoes' land. 'They were born on country and they'll die on country. She provides for all. You take that away from them and you got nothing.'

Dwyer had worked on the island as a ranger, on and off, since 1998. He thought the dingoes were bigger and stronger then, 'unless it was my own eyes,' he qualified. Dingoes scavenged at the dumps before they were closed in 1993.[4] 'They had cows way back,' Dwyer said. In the 19th century there were cattle stations on the island.[5] 'They had plenty,' he said. 'Too much.' He had heard how in the past big packs of dingoes waited till one of the brumby mares foaled and attacked foals when the mare was at her weakest. Though the Queensland government removed the brumbies in the early 2000s, reports of elusive brumby sightings persist and, mysteriously, horse was present in a minuscule proportion

---

4   Burns and Howard 2003, 704
5   Williams 2002, 77

of dingo scats collected between 2011 and 2014.[6] But horses do not make up a significant proportion of dingo diet on K'gari now. Dwyer thought the dingoes might be evolving or adapting to the space and resources now available on the island.

His attitude about working for Parks had changed too. When he was young he did not want to see rangers killing his wongaris. He thought it would be a conflict of interest for him to be a ranger, 'Being there watching people putting my companions down in front of my eyes. It made my blood boil. You don't do that. It's wrong. How do you know it's him?'

In 1999 he moved away down south. He worked his way back home and returned to the island in 2003. He was not there when Clinton Gage was killed in 2001 but he spoke about the sadness – for Clinton's family and for the dingoes because of the culls that went right down the island. His Butchulla countrymen and countrywomen felt just as passionate about the wongaris on K'gari as he did, he said. They were frightened of another big cull if there was another fatal attack.

After Gage's death Dwyer started looking at the island's dingo management from a different point of view and gradually, over the years, he told me, he learnt that QPWS dingo rangers were doing a good thing. The tags for so-called habituated dingoes provided identification so rangers could keep an eye on them. He thought it was a better strategy than before Parks had such processes in place.

'They are dealing with it the way our people dealt with it,' he explained. But he told me he thought the process should be quicker. 'Aggressive behaviour or nipping on the ankles,' he said, 'one strike and you're out.' The longer Parks left these dingoes, he explained, the more opportunity they would have to train their young to use the same tactics to gain easy access to food. He thought it was imperative to stop that cycle. 'If we don't stop that line, we're going to lose a lot of dingoes.'

The sound of passing traffic, chatter and the clink of teaspoons on coffee cups competed with Dywer's quiet voice. I leant in to listen more closely. 'Back when our ancestors were here, they dealt with rogue dingoes. They knew which ones were the bad ones, which ones might have come in and hurt their kids. They went and tracked them, great

---

6   Behrendorff et al. 2016, 2; 4, fig. 2

trackers they were, they tracked him down and put that dingo out of his misery so it didn't pass that style of tactics of killing kids and harming people through their generations.'

For Dwyer there was no question about which dingoes were dangerous. The habituated dingoes who weren't scared of people were the ones to be careful of. The timid ones who disappeared when you looked at them were the best ones.

He told me that as a Butchulla man, he questioned QPWS. 'What are you waiting for? Are you waiting for someone to get hurt? Let's deal with these dingoes now.' As a QPWS ranger he understood the monitoring system, the situation where the hierarchy needed to give the go-ahead to do things. But, he said, a cheeky dingo was like any animal on the mainland who got a thirst for killing birds or chickens. 'You don't deal with it inhumanely. You take it to the vet and deal with it there. You're not going to go, "Oh, he'll be right." He'll come back with all his pups and they'll do the same thing. They'll take a kid off you quicker than you can blink your eyes. They are cunning. Very clever hunting tactics … They go for your groin, areas where they can bleed you out. Chew you out while you're laying there. They don't muck around. Why muck around?'

Now, understanding better the strategy of how the government wanted to deal with them, he told me he could see himself becoming more involved with dingoes as a ranger. Not all the Butchulla community felt the same way, he said. He put it down to education, to wanting to know. Some didn't care, some did. 'We have these special meetings to try and get the community involved, to come along and listen, and be involved. More strength.'

He explained that the work of mediating was not easy. Ninety to 100 years ago, dingoes and people were companions. It was hard for people to hear that now dingoes are all rogues. Not that rogue dingoes were bad, he said, just that now they were on their own, wandering dingoes.

To Butchulla people who still wanted to have a camp-dog relationship with dingoes, he was realistic. 'Your home is in a house on the mainland. You're not constantly over there and that wongari's not going to sit there and wait for you to come back on the barge so you can walk with him through the bush. They've got to live their style of life there now. We've lost that companionship. They're in our hearts, that's their country now.'

He told me he found it hard to educate some of the elders, but they liked it when he listened, when he went away, worked on something, came back and gave them an answer. Sometimes, when he knew enough, he could give them answers straight away. He was constantly involved in the feedback process, which, it seems, became a loop. When he asked the elders for their opinion, 'Which way do you reckon we should go with this?', they put the question back to him: 'What do you think?'

He got on well with one of the dingo rangers who understood and respected Butchulla culture. With a Butchulla men's group he discussed dingo management, 'how we should be going about these things, what negotiations can we do'. People talked to him about feeding stations but he thought that would only make the dingoes bigger, stronger and worse. They would rely on people for food but they would not be pets.

He said that he liked talking to kids best. Visiting schools, talking about country, asking them for ideas about how to make things better. 'What would you do? How would you go about it? What do you think of this? How should we change it? What's a better strategy?' Children, he said, were our future. He loved how they would say how they were feeling, and their honesty about dingo interactions on the island. 'They'll tell you Mum and Dad left the bag out overnight and the rubbish got ripped out by the dingo.'

He felt that a lot of people who visited the island with their children assumed they knew the drill. But it went beyond sticking together. When they were camping or fishing for tailor on the beach, Dwyer said, they needed to move the dingoes along. They needed to understand that if they didn't move a dingo along it would be there looking at the next family's kids who, in a dingo's eyes, were the same height as their prey, the swamp wallaby.

It was a lot of work for a small number of staff to get the dingo-education messages out: move along, take your photos. Move them along. Stick together.

Language barriers didn't help. 'We're in their country,' he told people as part of the briefing. 'Let's respect them.'

When Dwyer came across dingoes in the course of his work, he moved them along. 'What are you doing here?' he'd say in a low urgent voice. 'Do you want to get in trouble?' Sometimes it was a nonchalant, 'Hello, how you going? Passing through? See you later, bud.' Sometimes he yarned with them in lingo. 'Don't want you to be habituated. Keep going,' he'd whisper. 'You don't want to hang around here.'

'Do they get it?' I asked.

'Oh, some of them,' he started before finishing with comic timing, 'mm, no.'

'Even though they look like they understand everything?'

'Some of them do listen,' he said. 'Move 'em along. You keep a distance, we'll look after you. Otherwise, you know, there's gonna be consequences.'

Dwyer explained that when he visited the island with his own children, he stayed with his kids and did what they wanted to do. If they wanted to go fishing, he went fishing. They didn't wander far from him because they knew the wongaris were dangerous. A wongari could reach a child alone on the beach 60 metres away before a human. To make rangers' jobs more difficult, he told me, holiday seasons coincide with important seasons in the dingoes' life cycle. 'Easter time, that's breeding season. That's holidays, a peak period, a high population of people. The dingoes are starting to run with packs. This is when we pick up our compliance, our Be Dingo Safe messaging.'

A car accelerated around the corner, its revs and gear changes a series of mechanical crescendoes that drowned out Dwyer's voice and aurally expressed surging hormones, sexual energy and holiday excitement.

'September holidays you've got the whelping season, when they're teaching their pups.' During September the young pups, born during winter, start to venture out, explore their territory, observe their adult pack members foraging and look for their own food. 'August and September are when the tailor fish run,' Dwyer explained. People fish for them along K'gari's eastern beach. 'Everyone's putting fish offal back into the ground or throwing it willy-nilly anywhere,' he said. 'Easy access for dingoes. And people say, "You gotta make them work." Which is actually true. You're burying these frames at the low-water mark so you get the high tide coming over them and washing them away. Other marine life deals with that or eagles'll come and snaffle them up.'

A supermarket trolley clattered over a speed bump in the car park.

Despite the rule that fish frames, which are the bones and unused parts of the fish after it has been filleted, need to be buried in a hole at least fifty-centimetres deep just below the high-water mark, Dwyer said some people were too lazy. I knew from my talks with the rangers that dingoes dig up these fish frames. Dwyer told me how rangers were constantly picking up people's refuse, talking to people, trying to enforce compliance. 'You've got to deal with some yobbos who think you're the fun police when you're writing out tickets for everybody. It's hard to be a ranger, especially when you've got to go and clean their toilets after a good talking-to-them. You've got to deal with all that rum and tailor fish.'

Dwyer had never been afraid of a dingo. He felt connected to them. When a dingo had been hit by a car or killed, or euthanased, by humans' hands, he said, 'I send them home country way, spiritual way. I smoke them and talk to the smoke, send their spirit home. Because to me they're walking in limbo land because they don't know what they've done.'

If a dingo had been killed he asked the rangers to ring him up, or, if he wasn't there, to talk to another Indigenous ranger. He needed them to bring the body of the wongari to him before they did anything to it. It was a problem when Parks needed the body for a necropsy. 'We tell them flat out. We want to deal with it our way.' It was important to release the wongari's trapped spirit first, before necropsies were conducted. 'I like it when they come up and ask,' he said.

For the ceremony, he lit a fire and sent a message in the smoke. He talked with the smoke and sent the wongari's spirit and the message. Once the spirit was gone, QPWS could do whatever they wanted with the wongari's shell. He didn't care if, at first, some rangers hadn't liked it. The elders were happy he was doing it the Butchulla way, 'our way', the proper way. 'The ones that get put down don't know what they've done,' he said. 'All they're doing is going for a feed.'

There was plenty of food on the island, Dwyer explained: 'Marine life – turtles, dugongs, whales, dolphins, birds. Swamp wallaby, lizards, turtle eggs.' He'd seen plenty of swamp wallabies. 'Swamp wallabies will go in the water and try and drown the dingo. Dingoes end up getting them drowned and drag them out.' He'd seen dingoes take an echidna out in the ocean, drown it and chew it up.

But he described how the island dingoes do not have much space. Mainland dingoes' territories might extend for 400 kilometres; K'gari is only about 120 kilometres long and twenty-three kilometres wide at its widest point. Survival was competitive. Of a litter of nine or ten, only three or four survived, Dwyer said. 'The rest get shunned out of their group and roam, and look for an area to move in. They're constantly moved out from other dingoes that own that bit there. Some of them travel from one end of the island to the other searching for their place, where they can make their own clan. Sometimes they fight to the death to get that.'

The traffic growled by as Dwyer spoke softly about dingoes attacking other dingoes in fights over territory. 'I've seen some action over there – four or five dingoes into it.' He hadn't seen the death; they went off into the bush. You could tell a dingo kill, he explained, because they went for the throat or the neck, used the power of their jaws to break their opponent's neck or back. 'They don't eat each other,' he said. 'They just kill to kill, for survival, and leave them where they are. Anything else they kill they kill to eat.'

'The old fella, the dominant male, might get too old, and the young fellas would have a go at him any chance they get,' he explained. 'The alpha females do the same thing.' Dwyer thought the males and females fought differently. I'd heard elsewhere that captive male and female dingoes fought differently: the males made displays of aggression and bit the thick fur around their opponent's throat; the females fought harder, with more will to hurt and maim. They went for their opponent's soft belly flesh and thighs.[7] Dwyer reckoned males and females howled differently too. If he sat down and listened he could hear the different pitches, different strength in the howls. He encouraged his children to listen to country. 'Clear their minds. Like a meditation. Reconnect. If you listen to the dingoes, one's howling out and the other one's over there and you can pick out which is the male and the female if you're clever enough. Our people had a lot of time for listening and thinking.'

He asked me whether I'd seen any dingoes on the island with the rangers the day before. The dingo population would be at its highest.

---

7   Watson 2015

'It would be at capacity now,' he said. 'They've got all the pups running around now.'

I told him about the paw prints around the hole with the rubbish. The paws of the adult males were over fifty-five millimetres long and wide, he told me. Females' and juveniles' paws were smaller.

He loved the way they ran in packs. 'They're playful.' Watching from a distance he could see they loved their families, kept them close until they were old enough to move them along. They were adapting; they were struggling, he said. He told me that some of them – the line of dingoes on the west side, in the middle of the island – stayed out of the way.

He thought the fences around resorts and camping zones were an improvement. He described it as keeping the so-called problem child in the fenced area and keeping dingoes safe outside the fence. 'You'll never go back to the way it was, used to be. Unfortunately that's the way it's going.'

We talked some more about rangers' powers to remove domestic dogs and cats from the island – just not on freehold land. So some people had dogs who never seemed to die; their owners alleged they had been on the island since before the rules changed in the early 1980s. Dingoes removed domestic dogs too. Dwyer had a mate, a timber cutter, who took his pet dog to the island. 'It was gone within three or four days. It ran away and never came back.' He told his mate his dog probably came out the other end of the dingoes the next day.

He hated the thought of dingoes breeding with domestic dogs. 'You might as well wipe them out. Mongrel breeds are the worst. They just kill and leave – anything and everything. Sheep, the lot.' He told me he hoped that there would be dingoes on the island for a long time to come. 'They are our last pure strain. Why would you want to wipe them out?'

We were coming to the end of our hour and I started to thank him. 'I hope I said what you wanted to hear,' he said.

I didn't want to hear that dingoes and people could no longer be companions on K'gari, only in their hearts. I wanted to find a way to repair the bond that Dwyer said was broken. I asked him if he wanted to say anything else.

'I want to make sure they're happy, you know, the rangers are happy, the right things are happy.' He was taking a long view. 'I want our children and their children and the next generation of children to see these dingoes. Last thing we want to do is lose our dingoes. And lose a kid's life. That'll start ...' I guessed he was going to say the killing of a lot of dingoes but he left it unsaid, and continued, 'which would be sad. I don't want to lose the dingoes.'

I asked him about the grief people living on and visiting K'gari must feel, because they wanted a connection with dingoes.

'They do,' he answered. People wanted to feed them, fatten them up, but he told them the dingoes' build was skinny, like a greyhound, that there was plenty for them on the island.

'Is it about something else other than feeding them?' I asked. 'They're oriented toward food, but is the relationship, the grief, about that companionship?'

'I still feel like they're my companions,' he began. At another point he had said they were not his companions. In the circumstances, I didn't think it was inconsistent. Ensuring dingoes kept their distance was his way of caring for them. 'Poor things,' he said. 'You know, I feel sorry for them.'

He told me about the past. Before colonisation the vegetation on the island was not as thick, except in the rainforest areas. The Butchulla did what Dwyer called 'mosaic burns' to keep the country open and accessible. 'Easy access, easy walks, easy feed, the clan line,' as he put it. The dingoes, he said, 'had to adapt back to the way they used to be back when our people were roaming the country. Once upon an island, once upon a time, they followed us, our mob over.'

He told me he wasn't clued in to the dingo dreaming on the island, yet, though a dingo whisperer had shown him a special dingo site, a place where they had managed to trap a female dingo they were after, and now he always paid his respects at that place. He told me about Aboriginal tribes on the mainland whose dreaming tale was the dingo. In his eyes dingoes had 'been here since Aboriginals been here, been part of us since day dot. Some people say they walked down, come here. Because they're unique.'

From a non-Indigenous perspective, no one really knows how dingoes arrived in Australia. One theory is that South-East Asian seafarers, hunter-gatherers from south Sulawesi,[8] brought them as domesticated or semi-domesticated animals to the north of Australia

---

8   Filios and Tacon 2016

about 3000 to 5000 years ago. From there they formed commensal relationships with Aboriginal people and 'colonised' the whole continent. The oldest remains archaeologists have dated indicate that dingoes were present on the Nullarbor Plain in southern Australia between 3348 and 3081 years ago.[9]

Genetic studies[10] take us back thousands of years, possibly tens of thousands of years, past glacial maximums and the inundation of land bridges, to illuminate prehistoric human migrations and canid domestication.[11] One hypothesis, supported by dingoes' genetic relationship to the New Guinea singing dog, is that they walked across a land bridge between Australia and New Guinea 6000 to 8000 years ago.[12] Recent genetic research suggests dingoes may have evolved on the ancient continent of Sahul over 20,000 years ago, and that the dingoes of K'gari are unique in the amount of genetic material they share with the New Guinea singing dog.[13]

According to genetic research the dingo's closest relative is the New Guinea singing dog, I told Dwyer. He told me that when the water was low between Cape York and New Guinea there were caves with artwork in them, all the way over. I didn't ask about the special dingo dreaming place on the island. Of course now I wish I had. But I didn't want to ask what I have no right to know. Instead I imagine a hospitable place, a soft place, from a dingo's point of view: a ridge with good sight lines; scrub and logs nearby to take cover in; shade to rest in; a clearing where the young pups play.

---

9   Balme et al. 2018
10   Ardalan et al. 2012; Oskarsson et al. 2011; Pang et al. 2009; Savolainen 2004
11   Pierotti and Fogg 2017
12   Cairns et al. 2017
13   Cairns and Wilton 2016

12
Over and over

The world she lives in is not mine.
   Helen Macdonald, *H is for Hawk*, 2014

The next day I embarked on a two-day Fraser Island safari tour. The first safety message delivered by Brett,[1] one of our guides, was about wearing seatbelts. The second was about dingoes. 'If you see a dingo, do not crouch down. Remain upright. Take only photos, walk back to the group. They are native here on Fraser Island and they are dangerous.'

The all-terrain bus climbed up the hill from the Kingfisher Bay Resort Village on the sealed road and careered down the track on the other side over rollercoaster moguls. Sitting at the front, I was alert – half thrilled, half apprehensive – to the way Brett manoeuvred the vehicle down the hill so wildly; the engine seemed excited too, whirring with what sounded to me like high revs. Perhaps tourists are foolish to trust the expertise of their guides. Or perhaps that is one of the appeals of being a tourist – childlike relinquishing of decision-making, curtailment of agency, simple trust in someone else's authority.

I was there to be a credulous tourist. Six months ago I had booked the tour as a way of getting to the island because K'gari is not an accessible place for someone who doesn't drive a four-wheel drive. I

---

1   A pseudonym has been used here and subsequently.

wanted to be as dispassionate as I could be, to hear and see different perspectives. I also just wanted to talk about dingoes. I thought of the tour as research. I had no idea what an amazing sensory experience it would be.

Our bus ploughed along the soft-sand track that in some places, under the wheels of the 70,000-plus vehicles that Brett said visited K'gari every year, not to mention those that stay on the island, had been etched a couple of metres below the level of the surrounding bush. He related the island's geological history: over millennia each beautiful elliptical grain of sand that makes up K'gari was brought in on the wind from the south or rolled in on southerly swells and was then overlaid with bush. Different sand dunes host different flora: drier sand under wallum scrub and damper sand under rainforest. He told us about K'gari's timber and where it was used – karri for ship's masts; hoop and slash pine for house frames; satinay for widening the Suez Canal in the 1920s and for rebuilding the London docks after 1945.

I can't remember the native species he talked about, apart from the last: dingoes who, he said, came to Australia with Asian seafarers as food. The food theme continued when, explicitly refusing to assign causation, Brett said he'd tell us two things about dingo attacks on the island. The last dump on Fraser Island was closed in December 1993. The first reported dingo attack, he claimed, occurred in January 1994. The dump at Uluru, he said, was closed two months before Azaria Chamberlain was taken.

The dingo information sign near the boat ramp at River Heads, embarkation point for the barge that had taken our tour group to the island, warned about the problematic relationship dingoes have with food, especially humans' food: 'Feeding dingoes is dangerous – for you and for them' and 'Encouraging dingoes to approach you increases your chances of injury'. It showed a photograph and a drawing of a dingo with its lower teeth bared, and a photograph of a dingo tugging at a plastic bag near an overturned esky at a campsite.

The pictures illustrated for tourists the foundation premise of the current Fraser Island Dingo Conservation and Risk Management Strategy:[2] if people provide food to dingoes, those dingoes become

---

2   Ecosure 2013

habituated and lose their allegedly natural fear of humans. The rangers had told me that these habituated dingoes are more likely to attack people. Feeding dingoes, the sign warned, increased people's chance of injury, and, in a catch-22 for dingoes, it was also dangerous for dingoes because QPWS staff kill dingoes who are too interested in humans.

The public was asked to collect data, but only one form of data:

Queensland Parks and Wildlife Service is continually monitoring Fraser's dingo population. Please help this process by reporting any negative encounters with dingoes to Rangers. Take note of the features of any threatening animal to aid in completing a Dingo Incident Report Form.

To facilitate reporting, there was a drawing of a dingo with identifying features marked: 'Body condition: ribs and hips well covered or obvious', 'Ears: erect or drooping', 'Ear tags: what colour, which ear', 'Injuries: scars, ragged or torn ears', 'Tail: limp, kinked, part tail missing', 'Tail tip: white or ginger', 'White feet and socks: extent above paws'.

Beside the sign's final message – 'Never encourage or excite dingoes. They can become aggressive and attack you. Dingoes that become a serious threat to people must be humanely destroyed. Please let them grow wild' – was a picture of a dingo pup, sitting, big ears pricked, looking pensively, attentively, at the viewer with dark triangular eyes as though it was the most natural thing in the world for a dingo to be interested in a person. But I had learnt that on K'gari 'wild' means afraid of people.

Maybe they were all wild dingoes at Lake McKenzie the day our tour group visited because we didn't see any. The picnic area between the car park and the lake was fenced. A sign informed us that it was an offence under the *Nature Conservation Act 1992* (Qld) to prepare or consume food outside the fenced picnic area. Food was prohibited on the lake shore.

After my swim I sat in the shallow water on the shore with my knees raised and exfoliated my feet with the lake's fine white sand. Little brown fish, shaped like leaves, swam toward me and hung around my legs and feet. They glided under my calves and rested in the shade of my legs. One moved over my foot so closely I could feel the gentle displacement of the water. They steered with their tails, oriented their heads toward me and I could see their bulgy eyes. It felt almost like affection, interest definitely. They were so unguarded they were tame. Were they young? Would you call them habituated?

When I asked our tour guides about them they explained that they were guppies (Western carp gudgeon; *Hypseleotris klunzingeri*) who feed on dead skin cells. What kind of benign wilderness is this, where animals are so unthreatened they approach humans? Nothing could be more unlike horrific stories of predation than this experience of being eaten by an animal. Guppies can make their unobtrusive approach to a human unnoticed. But dingoes are different.

At Central Station, formerly a centre of logging operations, was a 'Dingo-Aware! storage locker', a metal cage built on a wooden platform about one metre off the ground. In bold type, visitors were exhorted to lock up 'all iceboxes, items containing food, items that smell like food'. The helpful list of things that people might forget to lock up – 'small plastic food containers, bottles of oil and sauce, butter, wine casks' – read like an exotic wish list for an imaginary dingoes' picnic.

I didn't count the number of people on our tour – maybe about thirty. Most of them were from overseas – English, Swiss, French. Our group, along with several other tour groups, ate lunch in the dining room at the Eurong Beach Resort. After lunch our bus drove us north along the eastern beach. An English woman joined me on the seat behind the driver. She wanted to talk. She wanted to know how much I'd paid for this tour. She wanted to know what sights to see around Sydney. She wanted her money's worth out of her trip. As the bus rose and fell over the undulations on the beach I scribbled illegible words in my notebook.

Other buses and four-wheel drives zoomed along the beach too. It is gazetted as a road, our guides told us. Rules of the road apply. There is a speed limit, the wearing of seatbelts is compulsory and, of course, drink-driving is prohibited. We saw a police vehicle parked up on the dune. A few months before, Bold was the suspected 'thief' of a bag of breathalyser tubes. Rangers saw him nearby after an unidentified dingo had run into the dunes with a bag that contained rubbish including breath-test tubes and two empty tuna cans, which had been sitting on the ground near the police vehicle while police officers were conducting random breath tests. The bag was never found.

Our itinerary – how far north we could drive up the beach at what time – and the itineraries of all the other tour buses were dictated by the tides. We all stopped at the Maheno for the same fifteen minutes, but that did not diminish the wreck's rusting fascination. Most of our tour group scrambled up the rocky, grassy track to the top of Indian Head, an ancient outcrop of volcanic rock around which the sands of K'gari had gathered over thousands of years. From there, perhaps a couple of hundred metres above the beach, we could look north across the green-blue ocean and breaking waves to Waddy Point. To the south, the four-wheel drives on the beach looked like little black oblong insects. I paused during the short climb to watch an eagle circling high above the casuarinas. The joy it gave me to see how effortlessly the eagle rode the wind was tangled up with my wonder about what it was seeing and the sense that, if I could just watch it for long enough, I would learn something. From the summit of the headland, looking down at the water, our other guide Grant[3] pointed out five sharks. Tiger sharks, he told us. It was unusual to see that many together. They looked small from that height but they weren't, Grant said.

It must have been during my first visit to K'gari in the late nineties that I heard a story about a massacre on Indian Head: how, during colonisation, settlers and Native Police chased the Butchulla to the rocky outcrop and from there they jumped into the sea. I can't remember who told me this story. It doesn't appear in internet versions of island history. Later, at home at my desk, I read that the Butchulla name for Indian Head is Takky Wooroo; it is an important ceremonial place – a lookout, a story place, a sacred area reserved for men and a memorial site for the massacre I had heard about.[4] When I took the tour I didn't know that to climb Takky Wooroo is to show disrespect for this spiritual place.

It was sunset by the time our tour reached Champagne Pools. Lovers kissed each other in the golden light while I watched little fish swim around a rock in a shallow pool. By 5.45 p.m. we were driving

---

3   A pseudonym has been used here and subsequently.
4   Weisse and Ross 2017

back south at 80 kilometres an hour through the immense horizontal tunnel of the eastern beach: breakers on our left; the dune, which my eyes scoured ineffectually for dingoes, on our right; wet sand in front reflecting a wide sky action-packed with clouds. As we travelled, the expanse of beach we were heading into disappeared into the mist of dusk. I didn't see a dingo. I was thinking about Bold and his brother.

Near Eurong, a Swiss tourist at the back of the bus said, 'Dingo!' A pup was on the beach. Her mother climbed the dune. The tour guides knew the matriarch of the Eurong pack.

Brett swung the bus around and opened the door so we could see 08Purple lying on the dune taking in the evening breeze. We all stayed on the bus and people took turns to come down to the door in twos and threes to photograph her. I knew I had to give other people a turn at the door, but I was mesmerised. Her beautiful head rested on her white paws and she gazed serenely out to sea with bewitching dark-rimmed eyes. She looked confident and content. Her pinkish-coloured ear tag was bright in the fading light.

Not far away her dark, gangly daughter sat on the dune and watched us. I knew that she was one of the 2015 pups, about four months old. She came down and stood on the sand, her long front legs slightly splayed. I don't think she knew how vulnerable she was, how controversial it was for her to be standing on the beach between her mother and the bus, how her behaviour would be noticed.

Before we drove over the cattle grid into Eurong we saw another pup on the beach. A tourist commented on how skinny the dingoes looked and Grant reassured us that dingoes are naturally lean.

My room that night was on the ground floor in the two-storey white, besser-block 'Tradewinds' wing of the Eurong Beach Resort, a few hundred metres away from the two-storey weatherboard Eurong Beach Bar on the other side of the village. On the western side of the bar was a covered drinking area and on the eastern – beach – side was a swimming pool. East of the swimming pool was a track, called 'the Esplanade', which led to Second Valley. From the watered green lawn around the bar grew coconut palms, Norfolk Island pines, pandanus, banksia and acacia. The dingo-deterrent fence ran in a swampy dip east of the Esplanade, before the dune that fronted the beach rose up.

The fence around Eurong and Second Valley did not deter Bold. In early July 2015 two children at the QPWS accommodation at Eurong saw him crawl under the vehicle gate to get inside the fenced area. Interaction reports record him being inside the fence three times in July and August. The evening before the third 2015 State of Origin rugby league match, a dingo suspected of being Bold turned up to a barbecue in Second Valley and wandered off without showing aggression, visited the Eurong bakery, walked behind the resort pool, and entered a ground-floor room at the resort before being enticed out by its occupant, a tag-along tour leader who, on the report, said he could not remember who he worked for or how he got the dingo out of his room.

The next day two dingoes were spotted inside the fence. Bold was suspected of approaching people at the Eurong pool and at Second Valley. I could imagine him trotting up and down the Esplanade between Eurong and Second Valley, not wanting to get his paws wet and risk snakebite in the swampy bush outside the fence, visiting the places people were, interested in what they were interested in.

On 8 July, while Queensland trounced New South Wales fifty-two to six in the third State of Origin game, a dingo, suspected to be Bold, was hanging around the Eurong Beach Bar. Interaction reports record that he followed four separate parties from the bar that night. The first person headed out of the bar to the resort reception area, and clapped loudly and yelled at the dingo, who stayed about two metres away and started propping and jumping from side to side. When a ranger approached and started walking to the front of the resort, the dingo followed, but he would not be lured through an open gate. Next, the ranger headed west and the dingo followed him at a distance of five metres for 200 metres before he lost interest and went back toward the bar. He was then chased into the bush, but he reappeared and followed another person closely, within one metre, again propping and jumping backward and from side to side before once again losing interest and wandering off. Later he followed a group of four young people, at a distance of only half a metre, to their accommodation in the Tradewinds wing. He moved from side to side of another group of four adults as they made their way to Tradewinds from the bar. Each time he left after they went into their room.

Every year from 2006 to 2017, bar 2014, Queensland won the three-game State of Origin series against New South Wales. Rugby league does not define Queensland, but the Maroons' ascendancy and Queensland teams' never-say-die tenacity distinguish them from New South Wales teams of that era. I don't have much appetite for rugby league but I have been familiar with its repetitive routine – big men running in a line up a field, passing the ball and tackling – since I was very young. The language league commentators use, especially when two Queensland teams play against each other, is like the language people use about dingoes: teams play a 'sudden-death final'; one side is 'hungry'; the other is 'forced to blood' a new player; the teams 'went at it', they 'ripped into one another', had to 'fight our way back'. The players are 'superhuman', 'running doggedly' and one of them, like a dingo, is 'silky, the way that he moves'. It is easy to imagine Bold inhaling the beery passion of the Maroons' supporters at Eurong during the last 2015 State of Origin game, scenting the euphoria of Queensland's triumph, expressing his excitement, unable to contain his impulses, wanting to share the victory, emphatic that he too is a Queenslander.

* * *

The Eurong pack were up early the next morning: some of our tour party, who had been up at 4 a.m. to see the sunrise on the beach, saw five dingoes – they didn't know if they were pups or adults – walking from south to north as they returned to the resort. Ours was the first vehicle to park at the entrance to the Lake Wabby walking track. This stretch of beach was as nondescript as the rest of the long eastern beach but it is called a car park in the incident report that describes what happened there for ten minutes around 8.15 a.m. on 20 June when Bold approached two QPWS rangers who were going to clean the toilets. He dropped on his paws, crouched, sprang up and made short yaps and other noises but did not growl or snarl. The rangers labelled this behaviour as 'dominance testing'. When he opened his mouth and moved within half a metre of one of their legs they were concerned he would nip. He approached two tour groups and another vehicle before two more QPWS rangers arrived. When one of these rangers drove him away from the car park area by walking toward him with a post-hole shovel above his head, Bold ran backwards and forwards in front of the ranger. Then, with apparent wilful disregard for the fact that the beach is gazetted as a road, he lay down and rolled – close to the water in the 'traffic zone' – before trotting off southwards. On the Code D report detailing the morning's events three boxes were checked: 'dominant/submissive testing', 'dominant toward humans' and 'hunting tactics (with intent to test a response)'.

He did this sort of thing numerous times, usually only leaving – as he had walked away from me – of his own accord. Typically, and optimistically to my mind, he tended to leave one group of humans only to go and investigate another – not because someone had thrown sand and pumice at him, driven at him with a vehicle, waved a baseball bat at him, thrown bits of wood at him, chased him with a frypan, squirted liquid detergent at him, or gone at him with sticks and poly pipe and shovels. All these actions against Bold were described in the interaction reports and under the Fraser Island Dingo Conservation and Risk Management Strategy were perfectly acceptable, appropriate responses. Occasionally aggression and harassment warded him off; often it only seemed to excite and attract him.

Yet he did not nip a person until 23 July, after the radio-tracking collar had been fitted around his neck. This was one of the incidents Novak and I had discussed. Bold surprised a woman on the beach who was photographing her partner driving their four-wheel drive across a creek. She ran and he nipped her. When she stopped running he stopped. She walked to her vehicle, photographed Bold, rang QPWS from the barge landing at Hook Point and gave photos of the bite wounds, which did not require medical assistance, and of Bold to QPWS rangers at Rainbow Beach. Bold continued to procure food from camping zones and rip tents. Dangerous dingo signs were erected in the places he visited.

I didn't pay much attention to the parking area for the Lake Wabby walking track, or to the toilets – our guides specifically didn't recommend them. While my heart thumped with the exertion of walking uphill on soft sand over the dune to Lake Wabby I chatted with our tour guide Brett about dingoes. He, like others, thought that habituated dingoes had to be killed but he expressed doubt about how many dingoes were left on the island, how long they would be there.

I gleaned small subversive messages from possibly insignificant signs: when Brett told us the Butchulla word for dingo, he used *wat'dha*, the word for tame dingoes who live with people. Was this choice of words a subtle critique of the rule that they must all be rogue dingoes, *wongari*, now? Or was it the only word he knew?

Further north, at the upstream end of the boardwalk at Eli Creek, I stood under the green foliage watching the clear, thigh-deep water flow underneath me. A man arrived with a tripod and I moved to allow him to put it down. Immediately I regretted moving because, I thought, I had as much right to be there as he did. He set up to film a magnificent spider on its web. Along the boardwalk a Kingfisher Bay Resort Village ranger tried to catch a fly to put on the web so they could film the spider doing its thing. The filmmakers – a cameraman and a director – were from New York; they were making a documentary for the Smithsonian channel.

I went back to our tour bus – parked on the beach in a neat row with other tour buses, separate from the four-wheel drives – to put on my swimming costume and pick up an inner tube so I could float down the creek. The filmmakers and the KBRV ranger were still at work. I asked the ranger, who looked familiar, how long he had worked at Kingfisher Bay. It didn't take long to establish that he was Matthew,[5] a friend of my old flatmate Amanda, whom I hadn't seen since I visited her when she was working on the island seventeen years before. I was happy to hear about my old friend and about this serendipitous coincidence in such an out-of-the-way place.

We chatted and the cameraman talked about how he listened to Bob Dylan while he checked and catalogued the day's footage each evening. I wish I could remember which song he mentioned. Matthew was a Dylan fan too and demonstrated his knowledge by naming the album that particular song came from. Maybe he said it was from *Desire*, then he corrected himself; actually it was *Blood on the Tracks*. Or perhaps it was the other way around. I'm not sure but it seemed important then.

I floated down the creek on my inner tube, trailing my hands in the creek water and watching the bank glide past. When I put my head back my hair swirled around in the water and I could see the clouds and the

---

5   A pseudonym has been used here and subsequently.

sky framed by the tops of the trees. The slow motion of the creek and the view of the foliage and branches passing silently above me were deliriously gorgeous. No wonder Ophelia chose the river over rotten old Elsinore.

At 9.30 a.m. on New Year's Day 2015 a ten- or eleven-year-old boy floated down Eli Creek with his sibling. He drifted close to the southern bank and a young dingo jumped from the bank into the water and took hold of his hair. The children's mother, who was walking down the creek, ran toward them and screamed at the dingo, who let go. Two teenagers who were upstream chased the dingo, known as BWR for his blue, white and red tag, across the creek and to the north. It was allegedly BWR's third incident report for the morning. He was a seventeen-month-old male who was tagged at One Tree Rocks.

Earlier that morning, at 7.35 a.m., BWR was walking over the sand dune with another dingo before he ran across the creek, and circled and growled within one metre of a woman. After the ranger drove the woman back to her car in the ranger's vehicle, BWR stayed in the creek and hung around within five metres of other tourists. A photograph on the incident report shows his tail swishing in the sun-laced water, which does not reach his belly. The sable fur on his back appears to be wet, glinting. After ten minutes he climbed out on the northern bank and headed south to rejoin his dingo companion, who had retreated to the dunes. At ten past eight a dingo – assumed to be BWR; the people involved thought the tag was red – jumped into the creek and leapt through the water upstream past a group of six people who were wading up the creek. He circled back and swam up to one man, sniffed his shorts and rubbed against his leg before leaving. BWR was killed by Parks later that day.

I think about him tugging the child's hair and wonder whether a dingo tugging at Ophelia's hair might have broken her reverie, interrupted the 'snatches of old tunes' she chanted as her skirts dragged her down. She might have saved herself from her 'muddy death'[6] in the 'old foul river'.[7] If only she'd been at Eli Creek.

Out on the beach Matthew gave me Amanda's phone number. He was interested in my fascination with dingoes. He wanted to know

---

6   Shakespeare n.d., *Hamlet*, Act IV, scene vii
7   Smith 1978, 128

whether I'd talked to 'the photographer'. He did not approve of Parkhurst but he thought the raid on her house was, maybe, taking it a bit too far. Like everyone I'd spoken to who worked on the island, he said that habituation made dingoes dangerous. But QPWS needed to publish the evidence to prove this assertion, I said. The evidence that habituated dingoes were aggressive, he said, lay in twenty years of experience. When I asked him whether humans treating dingoes aggressively might make them more aggressive he said, like others had, that dingoes are cowards. What the people who called them cowards didn't say was that an injured dingo, whether hurt in a fight with other dingoes, or in a hunt, or by humans, has less chance of survival. Unlike a domestic dog, if a dingo cannot hunt and scavenge, it will not be eating.

It was too windy on the beach for the drone the filmmakers wanted to use. I asked Matthew and the filmmakers to play a Bob Dylan song for the dingoes. '"Idiot Wind?"' I suggested, wanting to demonstrate that I knew some Bob Dylan songs too. It was an angry song with teeth in it. 'Or "Hard Rain's Gonna Fall"?' Thinking of Bold's mother.

But now it is 'Brownsville Girl' – an eleven-minute epic with gospel backing vocals, tex-mex horns as jangly as a dingo's gait and a story delivered in a voice with trust issues and perfect timing – that puts me in mind of Bold. There are lots of characters in this song, outlaws, a movie, a road trip, an old love affair, scorching sun, a crime, an alibi and teeth as pearly as a young dingo's. There's a line about some babies never learning, which applies to the people who tried to manage Bold as much as it applies to Bold.

Like the narrator of 'Brownsville Girl' I'm sure that sometimes Bold didn't mean to trespass, he just found himself over the line. His penultimate interaction report records how he stood at the screen door on the deck of one of the QPWS residences at Eurong looking in at the rangers as they were finishing dinner. He moved away when one of them stood up. I imagine him there, wraithlike, asserting silently, in his dingo way, his relationship with these humans.

13
With their whole gaze

Instinct,
from where I stand,
[...]
looks like love
      Carrie Tiffany, *Mateship with Birds*, 2012

I picture the three of them, mother 08Purple, and her two sons 08Red
and PuGY (Bold), loping down the beach, flowing forward and
meandering like Eli Creek flows and meanders to the sea. All one, in
a disparate way, they follow the tracks in the sand where the vehicles
have travelled. They crisscross the breeze, sniffing in long inhalations
the rich 'flow of information in the air'.[1] From many directions their
wet noses, out in front of their bodies, perceive smells of pollen, plants,
goannas and bandicoots; scents of fish, pipis, worms and seaweed;
smells of petrol, human sweat and bait fish emanating from plastic
buckets near where the anglers stand at the water's edge like black posts.

The three of them make shapes as they move: three heads, six
dilated nostrils. The scents they inhale pass across hundreds of millions
of smell receptors in their noses. Their sense of smell is so strong, so
sensitive, they can detect minuscule particles, scents so diluted that

---

1   Knox 2003, 136

they are imperceptible to humans and to the instruments we have invented. Humans do not have a vomeronasal organ either, which is inside a canid's nasal cavity under the bone separating their nostrils above the roof of their mouth. With it they process not smells but pheromones, which give them sensory input about sex and social life that goes straight to their amygdala and then their hypothalamus.

They lower their noses to the sand and inhale in short sniffs. In a special part of their nasal cavity they hold scents and analyse them while they continue to breathe. Underneath recent, strong smells they detect older, fainter smells. Through their noses they read who else has been here, how long ago and what direction they are travelling. Their world of smell is delineated in a different way from the sharp edges of visual and aural perception, and has a different relationship to time. Smells don't disappear or cease; they linger and fade. Every surface offers complex multilayered information with many plots and many players who, even though they may no longer be seen or heard, are still present through scent. Everything lives in these smells: experiences they know and experiences they are yet to have, days and nights that have passed, moons that grow fat as puppies and shine through the casuarinas, moons that starve on dark nights.

They traverse their territory with their noses and their paws. On the ground 08Purple, 08Red and PuGY leave their own unique scent from the apocrine glands on their pads and between their toes. They tell others who care to know they were here and enrich their story by leaving a scat or urinating.

Their heads are arrow tips. Three noses, six eyes. They make the points of a hungry triangle as they travel: isosceles, scalene, equilateral. They are insatiable. The anglers look at the horizon, at the channel in front of them, waiting for tailor, unaware of the dingoes with their heads low, their noses extended, waiting for their opportunity.

It is late July. The moon is growing bigger, like the 2015 pups, and 08Purple is the hungriest. Her teats are large because she is suckling. The day before yesterday she spent the whole day on the beach at Eurong waiting, asking. People clapped and yelled and waved a fishing rod at her and she moved a little way away. Then she came back. She stood with her muzzle low, her ears forward in front of a couple sitting on the beach drinking beers. She looks gentle, good humoured in the picture on the interaction report. Perhaps she enjoyed their exasperation. Attentive, watchful, her back legs splayed wide, her tail curled to her left over her back. White waves broke behind her. She could smell the beer residue in the empty bottle propped up in the sand between her and the young lovers.

That afternoon 08Red tried to steal a plastic bag full of bait. Anglers had left it on the beach with their bags and tackle box while they stood ten metres away in the water to fish. 08Red, ears askew, tail down, gingerly made off with it toward the dune. An off-duty ranger followed him in a car. When 08Red dropped the bag to eat its contents, the ranger chased him off, picked up the bait, returned it to the fishermen and told them to secure it. The ranger took a photograph of 08Red sniffing the spot where the bait had been: his nose almost touches the ground and his ears are forward. His pale eyebrows are raised and drawn to the centre of his wide forehead, and his dark, triangular-shaped eyes regard the person taking his photograph with what seems to me to be a calm perplexity.

08Red's nose and whiskers are better equipped than his eyes to explore the sand in front of him because his vision for close objects is not as good as a human's. He would not have been able to distinguish the same range of colours as the ranger or the anglers, and he would not have been able to see the same amount of detail at a distance as them. He would have seen the beach at Eurong, and the casuarinas behind the dune and the inhabitants of K'gari in yellows and blues. Nevertheless, although his gaze is directed toward the camera, the photographer would have been just one element in the broad scene he perceived. Like the ranger and the anglers on the beach, he has good binocular vision so he could judge perspective well, but with his eyes slightly on the side of his wide head, his field of vision, at perhaps 270 degrees, is wider than that of humans, whose forward-facing eyes on a flattish face take in only about 200 degrees.

That morning, perhaps attracted initially by their motion, PuGY walked then ran down from the dune with his hackles up toward two boys who were walking along the beach. He would have perceived movement with greater acuity than the humans he harassed because the light-catching cells on canids' retinas can process light more quickly than humans' retinas can. This efficiency at detecting movement helps dingoes to hunt; it meant that PuGY saw 'a bit more world in every second'[2] than the boys, who stopped walking and started yelling and waving their sticks at him. But their rage only seemed to intrigue him. Their father, who was about fifty metres away, ran over and threw a piece of rubbish at him. PuGY picked it up, dropped it and moved toward the boys again before he headed off southwards toward some motorbike riders and a bus.

The same morning he approached a woman and her three daughters as they walked along the beach near Eurong. He swung in behind the two eleven-year-old girls, who were about five metres in front of their mother and younger sister, and started bouncing around, lunging and trying to nip the backs of their legs, as though it was the smell of their fear that was making him dance. Two passers-by in vehicles helped. One man, a wormer, got out of his car and started throwing things at PuGY, which put him off momentarily. The second vehicle herded PuGY toward the dunes, so the girls and their mother could walk back to their car.

The next day PuGY hung around the Govi camping zone, south of Eurong. He was assumed to have been the dingo who left his teeth marks on an unattended esky there. That afternoon 08Purple was hanging around fishermen on the beachfront at Eurong. She approached one group, lay down and watched them from a distance of about fifteen or twenty metres before moving to another.

The moon was getting bigger, waxing gibbous. 08Purple and PuGY were always hungry, waiting a little closer to the fishermen the next afternoon when a ranger's vehicle appeared. Its occupants exchanged words with the fishermen about keeping dingoes at bay, and moved mother and son on by yelling at them and driving at them. The dingoes moved south down the beach to the main entrance to Eurong.

A white LandCruiser drove out of Eurong and stopped. Its driver got out and crouched down two metres away from 08Purple to take a photo of her. The rangers drove up fast and straight as a bullet, moved the dingoes further south and told the photographer not to get out of their car and kneel in a submissive way near a dingo. PuGY stopped to sniff the tailgate, driver's and passenger's door of a four-wheel drive

---

2   Horowitz 2010, 131

parked on the beach. He sniffed a bait bucket with its lid on. The rangers arrived in their vehicle to chase him away. 08Purple went into the dunes and came out eating something white. She moved south when the rangers drove at her, and she must have dropped what she was eating. When they got close the rangers discovered it was a small white bird. Sometimes humans and dingoes are interested in the same things, but we do different things with them.

For fifteen minutes the rangers' vehicle followed PuGY and 08Purple down the beach. Perhaps the dingoes trotted casually, their big triangular ears tuned in to the sounds of the QPWS ute. They can hear higher frequencies than humans, and tiny changes in pitch. They can rotate their ears around on their broad heads; they can direct one or both ears behind, and rotate them in tiny increments to catch noises either side of them and in front of them. Perhaps they ran, the southeasterly streaming over them and beside their ribs; under them, along their pelts; through them, in their gaping mouths, lifting their empty torsos, filling up their lungs. Paws thudding on the reassuring sand, back legs bounding, heartbeats louder than the waves and the sound of the engine behind them.

08Purple's jowls are probably empty when she returns to the den. Her pups might mewl, suck and chew at her teats, and knead her belly with their tiny claws and their soft-padded paws. She'll need to catch a wallaby.

Like most dingo pups, 08Purple's 2015 litter were born in winter – June, July or August – in a den that she had chosen and perhaps dug out, maybe in a hollow log near water on a hill. They were born with their eyes closed and they could not hear. They found 08Purple through touch and scent.

After about ten days, when their eyes opened and they started to be able to support their weight on their puppy legs, she might have moved them to another, bigger, den. She still suckled them, but left them with another member of the pack, an alloparent, such as PuGY or 08Red, when she went out to hunt. She and other pack members would come back and regurgitate food for the pups. From the age of about three weeks, the pups would have been able to eat the carcasses their mother, father or alloparents brought back to them. From four weeks they became more mobile and 08Purple might have moved them to another site by putting a carcass in a different location, a rendezvous site. The pups would have followed her there and waited while she and other pack members hunted. Adult dingoes continue to regurgitate food for pups until they are four months old or, on K'gari sometimes, six months old. The pups learn social and hunting skills from the pack and by the age of about nine months are capable of living independently.[3]

08Purple comes from a long line of Eurong mothers who survive however they can – under the verandas, at the dump, making the most of their opportunities on the beach – and who are also fast, intelligent hunters. A tour guide showed me a picture of 08Purple carrying a wallaby down the beach for her 2015 pups before they were four months old. She had drowned it in a puddle, he told me.

There is documentation of how another female dingo from the Eurong pack, perhaps 08Purple's grandmother, hunted, killed, shared and consumed a wallaby on the beach near Govi Creek, about twelve kilometres south of Eurong.

Just after 8 a.m. on 19 December 2012, observers saw a wallaby emerge from the scrub on the dune onto the beach. It was heading south, toward the people who recorded these events, and it was panting heavily and lying down. About 100 metres behind it came a dingo. The wallaby stopped at the waterline, turned around and saw the dingo, now only fifty metres away, before it jumped into deeper water. The dingo stood where the wallaby had been and watched its head as it appeared to paddle and struggle in two-metre-deep water and rough surf.

After a few minutes the wallaby returned to the beach. It was panting and rested on all limbs. When it reached the swash zone, the dingo lunged, barging the wallaby's shoulder and biting at its neck, pushing it into the water as she knocked it over. For short intervals she held the wallaby's head down, under the shallow water, as waves washed over its body. Each time the wallaby stopped moving she released it. The wallaby struggled and even managed to get back on its feet, and jump away parallel to the beach, but the dingo kept forcing it back into the water. For about five minutes the dingo lunged, bit, pulled and held the wallaby down in the water by the throat, releasing it only when it stopped moving, until, about ten minutes after it had first burst onto the beach, the wallaby moved no more.

---

3   Parkhurst 2010, 53–9

The hunter then turned her head to the beach and a juvenile female dingo (who was older than four months but younger than a year old), known to be her daughter, appeared. The hunter raised her tail and bared her teeth; her daughter lowered her head, tail and ears, and approached submissively, cautiously. Once the young dingo had reached the carcass, her mother let her feed while she lay down a couple of metres away, panting. After about five minutes she joined her daughter eating the abdominal area of the wallaby while little waves washed against its corpse. Mother and daughter alternated. They disembowelled the wallaby and the mother ate part of its intestines while the daughter lay ten metres away. They were still eating when observers left at about half past nine.[4]

The mother was born at Eurong in 2008, one of the 'dump dogs', and had her first litter in 2009. She had subsequent litters in 2010 and 2012. The daughter who shared the wallaby was from the 2012 litter, her mother's last. She was not seen after April 2013, and was presumed dead.[5]

Dingoes are death specialists. 'A beautiful but cunning animal' was how Lindy Chamberlain described them.[6] As Dingo Foundation president Les Harris wrote to the first inquest into Azaria Chamberlain's death, they have to 'practise their hunting skills on a daily basis and at a very high level of efficiency in order to survive.'[7] On K'gari they use their environment in innovative ways to help them hunt more efficiently.

For example, around 1 p.m. on 13 June 2013, on the beach near the walking track to Lake Wabby, a five-year-old male dingo carried an echidna by its rear left foot claws from the vegetation on the foreshore down to the water and dropped it in the swash zone. A juvenile female dingo was nearby. The water came in and covered the echidna while it rolled itself into a defensive ball with its quills out to protect itself. The dingo watched the echidna, pawed at it and dug a hole underneath it in the wet sand. He then manoeuvred the echidna into the hole, which filled up with water as the waves retreated. When the echidna raised its

4   Behrendorff 2018
5   Behrendorff 2018
6   ACA 2012
7   Bryson 1986, 222 (Les Harris quoted)

head and snout to breathe, the dingo tried to bite it. He put its front paw on the echidna to push it down into the hole.[8]

On land, dingoes have been observed to give up after several hours on echidnas that dig into the soil substrate and lodge there with their sharp quills to protect them. But on the beach, the echidna stopped moving after an hour. The dingo pulled it out of the hole by its snout, turned it over and ate the flesh from its soft abdominal area. After he walked away, the young female who had been sitting twenty metres away approached and licked what remained of the subcutaneous dorsal skin. The male had shown dominant behaviour toward her during the drowning of the echidna. They both stayed near the echidna's remains for a few minutes before leaving.[9]

* * *

08Purple was born around July 2013. In QPWS records, dingo ages are calculated from July. She was not yet a year old when she met and mated with PiYPi in autumn 2014. He was two years older, born in 2011, and he was a big dingo, almost 23 kilograms when he was tagged as a subadult (that is, between one and two years old) on 24 March 2013 at Winnam camping zone, north of Poyungan Rocks and south of the village of Happy Valley. What brought him fifteen kilometres south to Eurong? Who chose whom?

Humans rarely see wild or even free-ranging dingoes hunt, mate, nurture and fight with one another. I've been told that even captive dingoes are secretive about their mating.[10] Our knowledge of their lives is fragmented, partial, defined by our own interests. Our presence changes the picture, but the Eurong pack are used to that.

When 08Purple was tagged on 24 February 2015 about five kilometres south of Eurong, she weighed just over sixteen kilograms, hardly bigger than her offspring from her first litter. 08Red weighed 13.6 kilograms when he was tagged eight days earlier at Cornwells.

8   Behrendorff 2017
9   Behrendorff 2017
10   Watson 2015

PuGY was tagged at Cornwells too, but not until 28 March 2015, when he weighed seventeen kilograms.

08Purple and PiYPi have remained a stable breeding pair for years, producing litters in 2014, 2015 and 2016. In 2017 they had a litter of ten, five of whom were still at Eurong in May 2018.

08Purple might survive for many years, like another female dingo who, Linda Behrendorff told me, was believed to have been born in July 2001, or earlier, when she was tagged and microchipped 500 metres north of Eurong in February 2003. In March 2012 she was retagged in the Gabala camping area, about thirteen kilometres north of the original capture location, because her ear tag had delaminated, which made the colours on the tag difficult to see. But the inside of her ear tag and her microchip confirmed her identity.[11]

Between 2004 and 2014 she was photographed many times by people and on trail cameras around her territory north of Eurong – Cornwells, Gabala, One Tree Rocks and Poyungan. She didn't harass anglers or take food from campsites. Pictures show her feeding on dolphin and turtle carcasses. She was lactating in 2004. She was observed with pups in September 2006, September 2007, October 2008 and October 2011. It is possible she was PiYPi's mother. She was pregnant in June 2012. Over her long life people saw her with three different mates.

When Jennifer Parkhurst and I discussed this dingo, she did not believe that a K'gari dingo could live beyond the age of ten. They age more quickly on the island, she explained, because they are eating sand all the time and their teeth get ground down.[12] I believe Parkhurst. I believe Behrendorff too.

In mid 2014 on the eastern beach the allegedly thirteen-year-old female dingo was attacked and badly injured by two male dingoes from a neighbouring pack. She took cover under a vehicle until fishermen and rangers distracted her attackers and she got away. She ceded. She walked. In August she was seen about twenty kilometres north of Cornwells at Eli Creek. Later in August a trail camera caught her image – right ear still injured, pale muzzle down – more than twenty-five kilometres

---

11   Behrendorff and Allen 2016, 257
12   Parkhurst 2018

from Eli Creek at Awinya, on the north-western side of the island. However incapacitated she was by injury and age, she kept walking, as though walking was what she was made for. In October her partly mummified body was discovered 100 metres north of the Ngkala Rocks bypass, on the far north-east coast, more than twenty-five kilometres from Awinya and sixty kilometres from where she was originally tagged and microchipped. No cause of death could be determined. She was lying on her left side. Her right ear drooped stiffly forward over her eye. She was the colour of sand and sand had started to cover her legs. Wispy dune grass was creeping forward over her rump.[13]

---

13   Behrendorff 2015; Behrendorff and Allen 2016

14
Sore feet, tears and seeds

clay stops many a warrior's mouth, wild singer
Judith Wright, 'Trapped Dingo', 1944 [1946]

After the tour, just after take-off on the flight home to Sydney from Hervey Bay, I looked down at the estuaries, waterways, mangroves, sandbars and islands below, unable to distinguish mainland from island, land from water. The Mary and Susan Rivers mingled and flowed out into the strait between the coast and K'gari, making swirling patterns of turquoise, dark green and bright pale yellow. The colours were warmer than the cool ultramarine tones of my own New South Wales coast.

Cape Byron in New South Wales, at 153°38'20"E, is the easternmost point on the Australian mainland, but on maps K'gari sticks out to the north-east where the east coast of the mainland turns north-north-west, which gives the impression that Sandy Cape, the northern tip of the island, at 153°12'31.18"E, is further east. On maps, North West Cape in Western Australia and Sandy Cape look like two handles of an urn.

In May 1770, when Lieutenant James Cook and the crew of the *Endeavour* sailed north along the east coast of Australia, Cook thought K'gari was part of the mainland. He named it 'the Great Sandy Peninsula'. Through his telescope he observed people gathered on a rocky promontory, which he called Indian Head. Edward Armitage, an Irish-born timber-getter and skipper who arrived at Cooloola in the 1860s, recorded, translated and in 1923 published songs that offer a Butchulla perspective on the British explorers. One commemorates the passage of the *Endeavour* up the east coast of K'gari:

> These strangers, where are they going? Where are they trying to steer? They must be in that place Thoorvour [a dangerous shoal near Indian Head], it is true. See the smoke coming in from the sea. These men must be burying themselves like sand crabs. They disappeared like the smoke.[1]

Another song records the 1799 landing of Matthew Flinders at Watoomba Creek (now called Wathumba) on the north-western side of 'the Great Sandy Peninsula':

> Two times held up something and made loud noise[2] and smoke.
>> Their heads are like dingo tails.
>> The paddles are like wood shaped by the fire.[3]

Just as Cook named the Butchullas' country in terms familiar to him – 'Indian Head' – the Butchulla described the mariners, unfamiliar beings, in terms of animals who were familiar to them – sand crabs and dingoes. Armitage postulates that the Butchulla likened Cook's men to sand crabs burying themselves when the *Endeavour* disappeared over the horizon.[4]

But there is disagreement over the intriguing simile 'heads … like dingo tails'. In his gloss, Armitage maintains that the Butchulla, 'not knowing anything of clothing or head covers', thought that the mariners hats were part of their heads, and '[t]he tails were probably ribbons

1   Armitage 1923, 96
2   'Kong! Kong!' in Butchulla
3   Armitage 1923, 96
4   Armitage 1923, 96

streaming from the band'. He uses his own experience to verify this hypothesis when he writes, 'I often wore them'.[5]

Armitage's editor F.J. Watson disagrees:

> In the days of Cook and Flinders, the common head covering of seamen, when at sea, was a kerchief bound round the head and tied behind; moreover, the seamen in that day wore their hair in pigtails. The likening of the headdress to dingo tails was probably inspired by the fact that the adult male aborigine, when in 'full dress,' usually had his head adorned with a forehead band made of dingoes' tails.[6]

Dingo-tail headbands were worn by Indigenous people around Sydney in 1788.[7] I've seen nineteenth-century photographs showing both men and women wearing them around the Clarence River in New South Wales. They were also worn in the Torres Strait, Northern Territory and Western Australia and, I'd guess, other places. Dingo-fur hair ornaments as well as tails were worn.[8] But it is likely that neither Armitage's nor Watson's explanation approaches the depth and the breadth of what 'heads ... like dingo tails' means.

Archeological evidence from the south coast of New South Wales shows that tame dingoes have been part of Aboriginal people's lives for at least 1000 years.[9] Dingoes are deeply embedded in Indigenous knowledge systems across Australia.[10] The arrival in 1788 of the British at Warrane (or Sydney Cove) had a profound impact on all aspects of the lives of the Gadigal people, the traditional custodians of Sydney, and their dingoes. In an augury of what was to come, Watkin Tench describes an incident in which Governor Phillip's gamekeeper, a convict called John McIntire (or McEntire or McIntyre), shot a tame dingo belonging to local people:

---

5   Armitage 1923, 97
6   Armitage 1923, 97
7   Fowell 1788
8   Philip 2016; 2017b
9   Gollan 1984
10   Philip 2017a and b; Rose 2000; 2011

These animals [dingoes] are equally shy of us and attached to the natives. One of them is now in possession of the Governor, and tolerably well reconciled to his new master. As the Indians see the dislike of the dogs to us, they are sometimes mischievous enough to set them on single persons whom they chance to meet in the woods. A surly fellow was one day out shooting, when the natives attempted to divert themselves in this manner at his expense. The man bore the teazing and gnawing of the dog at his heels for some time, but apprehending at length, that his patience might embolden them to use still further liberties, he turned round and shot poor Dingo dead on the spot: the owners of him set off with the utmost Precipitation.[11]

European settlers arrived in Cooloola from the 1840s but Indigenous resistance was vehement and, for a time, successful; colonial control over the region was tenuous until the late 1850s.[12] During the frontier wars, K'gari was an Aboriginal stronghold, a sanctuary where the warriors who stole sheep and speared settlers retreated.[13] From 24 December 1851 to 3 January 1852, in an 'attempt to arrest blacks charged with felonies and murder, taking refuge on Fraser's Island',[14] notorious Native Police Commandant Frederick Walker along with a lieutenant, a sergeant-major, four local squatters, the captain and crew of a logging schooner and twenty-four Native Police troopers raided the island.

From 31 December to 2 January, according to Walker's report, 'Native Sergeants Edgar & Willy pursued blacks for 3 days to East side of Island – blacks took to sea'. Walker was 'too footsore to accompany men'. Half of the police were lame, too, and hardships such as rain, heat, mosquitoes and sandflies rated a mention,[15] but not dingoes. Walker's party brought two Aboriginal men, Durobberee and Periker, back to

---

11   Tench 1789, ch. 11
12   Evans and Walker 1977, 45–58
13   Bidwell et al. 1851
14   Walker 1851
15   Walker 1852

Maryborough to be tried for the 'murderous assault' against George Furber, one of Maryborough's founders, five years earlier in 1847.[16]

Walker's brief statement 'blacks took to the sea' may be a reference to the massacre at Takky Wooroo (Indian Head) that I had heard about. In some accounts, the people fleeing the punitive raiders were kept in the water 'as long as daylight or life lasted'.[17] Official colonial records do not reveal anything more about the massacre and twentieth-century archaeological surveys of Takky Wooroo have not uncovered physical evidence of human remains, yet Butchulla people have passed down accounts of the killing.[18] This knowledge persists, alive and plausible.

Historical accounts resonate with contemporary narratives of K'gari as a place of conflict and danger where the ongoing lethal control of dingoes is necessary. Here, history is not past; it is not over. As African-American writer James Baldwin puts it, 'the great force of history comes from the fact that we carry it within us, are unconsciously controlled by it in many ways, and history is literally *present* in all that we do'.[19]

Contact between Indigenous people, dingoes and settlers on K'gari and around Cooloola was not one event but an accumulation of interactions and incidents that began in the late eighteenth century and came to be as incessant as the waves breaking on K'gari's eastern beach; as recurrent as the birth of generation after generation of dingo pups every winter; as relentless as Bold, who could not be conditioned by all sorts of aggressive acts on the part of people to forgo his interest in humans, and who was killed for challenging a particular kind of human and government authority. The contact zone, according to historian Mary Louise Pratt, is a social space where 'cultures meet, clash, and grapple with each other, often in contexts of highly asymmetrical relations of power'.[20] On K'gari and around Cooloola, especially in the beginning, asymmetries of power were not all one way. Australian historian Inga Clendinnen describes 'the difficulty of generalisation':

---

16 Walker 1852
17 Weisse and Ross 2017, 155 (*Moreton Bay Courier* quoted)
18 Weisse and Ross 2017
19 Baldwin 1965, 47
20 Pratt 1991, 34

Clearly it took all types to make a colonial world ... there was a mosaic of different social relationships determined by local circumstance and particular personalities ... there was only one near constant: the insisted upon inequality between white, and every shade of black, and even that racial inequality was challenged by individuals.[21]

Nevertheless, as writer Alexis Wright points out, 'historical lies' and 'racist stories' have formed the basis for 'the physical control and psychological invasion of Aboriginal people' from 'the beginning of colonisation two centuries ago'.[22] Dingoes play their own parts in these narratives.

* * *

In winter 1836, before Europeans arrived to settle Cooloola, a group of shipwreck survivors spent six weeks living with Aboriginal families on K'gari and the Cooloola coast. One of them, a seaman called Harry Youlden, published an account of the shipwreck and its aftermath in the United States seventeen years later.[23] Like the song Armitage recorded, Youlden's description of how the locals dressed shows the importance of dingoes to people:

A small strap of dog-skin round the neck of the women, from which is suspended the basket which they use in gathering roots, and a narrow girdle the maidens wear till marriage, were all the dress we saw among them.[24]

The locals had no use for the fancy shirts or gold the castaways traded with them for fish. The clothing they 'tore ... into strips, and bound ... as fillets round their heads, or wore ... as girdles or bracelets, by way

21   Clendinnen 1999
22   Wright 2016
23   Youlden 1853
24   Williams 2002, 42 (Harry Youlden quoted)

of ornament' but soon threw away. The sovereigns they 'carelessly cast upon the ground as valueless'.[25] But they valued their dingo skins.

The castaways had left their brig, the *Stirling Castle*, breaking up on Swain reefs (north-east of present-day Rockhampton) on 21 May. After thirty-two days at sea they beached their unseaworthy longboat in rough waves north of Waddy Point on K'gari's east coast. They walked south along the beach and eventually stayed with the Butchulla, helping them fish when they could keep up with the locals as they travelled through bush that was, according to the Europeans' metaphor, 'so closely woven together ... it forms ... a massive wall of impenetrable wicker work'.[26]

In August 1836, the remaining seven survivors, including the captain's wife, Eliza Fraser, were found by British personnel and taken to the penal colony at Moreton Bay, the northernmost British settlement in Australia at the time.

In 1838 a London court reporter, John Curtis, who covered the Lord Mayor's Inquiry into Mrs Fraser's so-called captivity, published a book, *Shipwreck of the Stirling Castle*.[27] Curtis used oral and documentary testimony mainly from three *Stirling Castle* survivors – Eliza Fraser, John Baxter and Robert Darge – who all mention dingoes and their relationships with the Aboriginal inhabitants of K'gari. In these tellings, dingoes' dislike of the newcomers is one of their defining features.[28] Curtis's book is a rich and problematic source. By contemporary standards it is unreliable and racist. Particularly interesting is the way depictions of Aboriginal people and dingoes are entwined.

After their stranding, the shipwreck survivors took eight or nine days to trek 100 kilometres south along K'gari's eastern beach. They walked at night, often in the surf, so that the locals could not track their footsteps in the sand. But their efforts to avoid detection failed. Curtis denigrates Butchulla acuity by likening their keen sense of smell to that of their dingoes:

25   Williams 2002, 42 (Harry Youlden quoted)
26   Curtis 1838, 69
27   Curtis 1838; McNiven et al. 1998, 2
28   Curtis 1838, 58, 65, 224

This [walking in the surf] was certainly a prudent precaution, but they afterwards learned that it was a useless one, for such is the sensitiveness of the olfactory nerve of these barbarians, that they can scent the progress of Europeans as well as their quadruped brethren, the *blood-hounds*.[29]

As the castaways walked down the beach,

[t]he first thing which diverted their minds from the monotony of the scene was the traces of footsteps of animals, which they afterwards found to be those of kangaroo dogs.[30]

In a footnote Curtis explains that:

These dogs are very numerous, and although they run wild, they are frequently domesticated by the natives; and although they live among bipeds more savage than themselves, they seem to be endowed with a considerable share of instinct peculiar to their species, as by a little training they are taught to go out after kangaroos, which, when caught, they bring and lay down at the feet of their master. In size and appearance, they are not unlike a grey-hound, and naturally very inimical to white men – perhaps by tuition.[31]

Like their dingoes, when the Butchulla gave something to the castaways they put it on the ground rather than in the hands of the recipients:

The natives made a signal by friendly gestures, and held up something, which upon a nearer approach turned out to be a piece of a kangaroo, which they laid on the ground, as they refused to take any thing from, or give any thing into their hands.[32]

---

29   Curtis 1838, 57, emphasis in original
30   Curtis 1838, 57
31   Curtis 1838, 39
32   Curtis 1838, 39

So seaman Robert 'Big Bob' Darge competed with children and dingoes for his food:

> Darge also states with Baxter, that the natives never give any thing into the hand of their captives, but throw the article intended to be conveyed, at them, the miserable allowance of food not excepted. Sometimes he had only some fish-bones to make a dinner of; and when they were cast at him, the children used every endeavour to deprive him of his scanty morsel, and in this attempt, the dogs, which have also a great aversion to a white person, used to join, so that he often lost a portion of the food which had been intended for him.[33]

Curtis cannot relate that second mate Baxter was 'once attacked and bitten by a wild kangaroo dog' without also mentioning the dogs' 'decided antipathy to white men'.[34]

It is unclear whether it is Curtis or his informants who suspect that it is 'perhaps by tuition' that dingoes are 'naturally very inimical to white men'[35] but the preoccupation with dingoes' 'natural' as opposed to learnt or taught behaviour persists in the current Fraser Island Dingo Conservation and Risk Management Strategy's premise that 'wild' dingoes, who are 'naturally' wary of people, behave in the way that dingoes should behave and that 'habituated' dingoes, who have been fed by people, behave in a way that dingoes should not behave.[36] These distinctions rely on cultural assumptions about learnt and natural behaviours. The attendant judgments about whether these behaviours are good or bad, acceptable or unacceptable, relate not to the behaviour of dingoes but to the interests of the people who create the defining narratives.

The need to occupy and control territory, and fear, with its attendant shame, have occluded many settler Australians' ability to see Aboriginal people, and dingoes, clearly. Curtis's and his informants'

---

33   Curtis 1838, 224
34   Curtis 1838, 66
35   Curtis 1838, 58
36   Ecosure 2012, 78–80; 2013, 7

unfavourable comparisons of Aboriginal people with dingoes are emanations of 'the Australian psyche, its fear of the other',[37] but the effects of the 'sensationalising of racism'[38] are serious and far reaching, and serve a political agenda for colonial settlers, as Goenpul sociologist and activist Aileen Moreton-Robinson explains:

> The existence of those who can be defined as truly human requires the presence of others who are considered less human. The development of a white person's identity requires that they be defined against other 'less than human' beings whose presence enables and reinforces their superiority.[39]

* * *

As well as divisive, culturally constructed categories, seeds of different kinds of relationships exist among the *Stirling Castle* survivors' narratives. Curtis's informants' perceptions of mistreatment by Aboriginal people were directly proportional to their social status. The captain's wife, Eliza Fraser, and second mate and captain's nephew, John Baxter, were extremely critical while seamen Robert Darge and Harry Youlden acknowledged the hospitality they were offered and seemed to develop some respect for the locals. Darge attributed 'his suffering more to the severity of his labour, and being exposed naked to heavy rains, dense fogs, and furious blasts, than to acts of torture practised on his person by the natives'. He told Curtis that the reason Aboriginal people generally hated the British was 'the fact of their having been frequently and sometimes very wantonly fired upon by the soldiery and constabulary force connected with the colonial settlement'.[40]

After the castaways reached the southern tip of the island, the Butchulla ferried them in their canoes in ones and twos to the mainland, not that the Europeans knew they were crossing from an

---

37   Wright 2016
38   Wright 2016
39   Moreton-Robinson 2004, 76
40   Curtis 1838, 222–3

island to the mainland – they thought they were fording a river or a lake. Europeans did not find the sea passage between K'gari and the mainland till the 1840s. Three of the survivors, seamen Youlden and Darge and steward Joseph Corralis, continued walking south in an attempt to reach Moreton Bay. According to Youlden:

> We soon met natives, and ... they gave us food, and we remained with them three weeks, generally on the move; sometimes for many days with food in plenty, and for several together without any ... we joined an old man and two women going south. At a camp of about twenty natives they left us, and two young men undertook to guide us ... Selecting the native with the best lot of fish for my own particular host, I took modestly the most comfortable place by his fire, to the windward, away from smoke.
>
> He was evidently flattered by my choice, hospitably shared his fish with me, and we broiled and ate, and for lack of other mode of conversation, looked at each other pleasantly ... Their village was two miles off, and when, my feet having been pierced by the sharp stubble of burnt grass, I stood still from pain, my kind entertainer, evidently a man in authority ... took me on his shoulders, and carried me to the village.[41]

When the locals indicated that Youlden and his shipmates should move on and Corralis was distressed about getting lost,

> a generous young brave seized his spear, and volunteered to pilot us to the English settlements. We had not gone many miles, when, suffering excruciating agony from my bruised and battered feet, I told my companions they must go on and leave me ... [C]rawling, rather than walking, [I] at last reached the village, where the old men and women left in charge kindly received me.[42]

Youlden crawled – a gait more like a dingo's than a man's – and was received kindly. On 8 August, Darge and Corralis, 'black and perfectly

---

41 Williams 2002, 44–5 (Harry Youlden quoted)
42 Williams 2002, 45 (Harry Youlden quoted)

naked', encountered a British hunting party on Bribie Island.[43] Subsequently, on 17 August, convict John Graham, an indispensable member of the British rescue party, found Eliza Fraser camped with Aboriginal people at Lake Cootharaba (north of what is now Noosa). Her feet, too, were 'lacerated'[44] and she was 'quite unable to walk'.[45] Her rescuers 'were obliged to carry her in turns' to their boat for the return to Moreton Bay.[46] This apparent chivalry is undermined by Eliza Fraser's complaint to Captain Fyans, the commandant at Moreton Bay, that 'the white men she met treated her more harshly than the blacks'.[47]

* * *

In October 1859, twenty-three years after Eliza Fraser was carried to a British boat, the crew of the schooner *Coquette* carried another person, this time a girl called Mundi, estimated to be seven to ten years old, who might have been exercising the only resistance available to her by not walking. She was taken with her sister Coyeen, who was estimated to be fifteen to seventeen and had a sore toe.

William Sawyer, owner of the *Coquette*, describes the casual brutality of the abduction:

> Having observed the white children, a rush was made – they were seized and carried off – the tribe being panic-stricken and offering no resistance. Afterwards we travelled, it being fine moonlight, about ten miles to the southward, and camped for the night. At near midnight an alarm was raised that strange blacks were watching our movements; a volley of musketry was fired for intimidation, and we were not further molested. Next day we travelled about twenty-seven miles, the youngest girl having to be carried the greatest portion of the way. On the day following, the

43   Brown 1998, 23 (Charles Otter quoted)
44   Brown 1998, 25 (John Graham quoted)
45   Curtis 1838, 237
46   Curtis 1838, 187
47   Schaffer 1995, 93

bush having been fired by the natives, our walking was rendered very difficult, but we were able to regain the vessel.[48]

Captain Richard Arnold of the *Coquette* was also 'much fatigued' and 'footsore' by the time they made it back to the ship.[49]

The girls were described by Archibald Meston, who had been in charge of the reservation at Bogimbah on K'gari and Protector of Aborigines for southern Queensland, as 'aboriginal albinos whose mother, "War-ann-oong," had never even seen a white man'.[50] They were abducted because there were rumours that a white woman and child were living on K'gari. Sawyer claimed the two girls were survivors of the shipwreck of the *Sea Belle*, which had disappeared without trace after leaving Port Curtis, near Gladstone, in 1857 and whose passengers had included a woman with a broken toe and her two children.

To the Board of Inquiry appointed in Sydney, Captain Arnold offered a cryptic explanation:

> I have no means of knowing who the children are, but I believe they are white children ... The foot-print in the sand of the woman with the broken toe corresponds with the foot-print of the elder girl rescued. We observed it in the sand mixed with a hundred of others; but it was going in the contrary way to that which we were.[51]

I do not know how Arnold ascertained the girls' alleged European background from their footprints. I assume he said he believed that Coyeen's footprint corresponded with the print of the shipwrecked woman with the broken toe because he, Sawyer and his party wanted to claim the NSW government's £300 reward for bringing the shipwreck survivors to Sydney. I interpret his observation of Coyeen's footprints travelling in a different direction from other people's as an attempt at a form of literacy, which was practised with such expertise by the

48 Skinner 1974, 7 (William Sawyer quoted)
49 Skinner 1974, 9 (Richard Arnold quoted)
50 Meston 1905, 5
51 Skinner 1974, 8 (Richard Arnold quoted)

Aboriginal trackers who followed the prints of the dingo who took Azaria Chamberlain and the Butchulla men who travelled to Victoria to find Ned Kelly, and continues to be practised by people, dingoes and other animals who read tracks and spoor. Even though now literacy is equated with reading and writing letters and words, I do not think humans were ever 'pre-literate'. Humans and other animals have always read animal tracks and spoor, and patterns in the way animals, birds, fish and insects move. Moroccan writer Abdelfattah Kilito might have been partly joking when he suggested that animals actually wrote before humans, and that animals taught humans to read,[52] but I keep thinking about these other writers, and ways of writing, and I realise I have much to learn.

* * *

The presence of dingoes in the histories of convicts who absconded from Moreton Bay and lived with Aboriginal people is oblique, but it is there. Some of these escaped convicts, the lowest of the low in the British caste system, eventually became guides, interpreters and cultural mediators between settlers and Aboriginal people. On the shifting sand of K'gari and Cooloola, colonial hierarchies and certainties were unstable. For example, Irish-born John Graham, the hero of the search for the survivors of the wreck of the *Stirling Castle*, participated in the rescue because he was promised a pardon. He used the cultural and linguistic knowledge he had acquired living with Gubbi Gubbi (Kabi Kabi) people from 1827 to 1833 to negotiate the peaceful return of the castaways to the British.

James Davis, born in Scotland and known to Aboriginal people as Thurrum-Voi, meaning 'look quickly',[53] interpreted for the K'gari girls to the Board of Inquiry. He escaped from Moreton Bay in March 1829 and was only persuaded to return in May 1842 when explorers Andrew Petrie and Henry Stuart Russell found him and told him the penal settlement had been disbanded. While he lived with Ginginbarrah people at Cooloola, he accidentally killed his adoptive Aboriginal

---

52 Warner 2014 (Abdelfattah Kilito quoted)
53 Evans and Walker 1977, 93, n. 26 (Olga Miller quoted)

mother's dingo. According to his obituary, the way he avoided being killed in retribution was by giving his adoptive Aboriginal father, Pamby-Pamby, a 'merciless drubbing'.[54]

The two girls abducted from K'gari told Davis that 'their parents were on the island and were blacks' and that they wished to go back to them. Billy, an Aboriginal man who had helped the *Coquette's* crew find the girls and sailed with them to Sydney, confirmed that they were natives of the island.[55] But the Board determined the sisters should be sent to the Orphan School at Parramatta or the Destitute Children's Asylum.[56] Although Billy and other Aboriginal people from Cooloola asserted that the girls came from K'gari and there was clearly some level of official disbelief that they were European shipwreck survivors, Coyeen and Mundi's testimony was disregarded. Despite the promises made to their parents on the island when they were taken that they would be brought home,[57] according to a letter written to *The Queenslander* by Captain Arnold's son, Coyeen, or Kitty as she was called in Sydney, died not long after being brought to Sydney, and Mundi, or Maria, was sent into domestic service and died in a Sydney hospital in 1878.[58]

With hindsight it is astonishing how colonial authorities explained their observations with stories for which there was no evidence: the Sydney press claimed the girls' skin had been covered with a pigment, their noses had been broken and their lips disfigured – with the inference this was done on purpose by Aboriginal people; the girls could not remember English because of their ordeal, not because they had never heard it.[59] The colonists' blindness and deafness seem nonsensical and preposterous but there is a term for their treatment of Coyeen and Mundi: epistemic injustice. I am stuck on the notion of epistemic injustice, just as I am stuck with the idea that the past is not over. Epistemic injustice includes testimonial injustice, when a speaker

---

54 Anon. 1889, 14; Parkhurst 2015a, 38
55 Skinner 1974, 10 (James Davis quoted)
56 Skinner 1974, 6
57 Skinner 1974, 12 (Archibald Meston quoted)
58 Skinner 1974, 12 (W. Arnold cited)
59 Skinner 1974, 6, 11

is not believed because of prejudice, and hermeneutical injustice, which occurs when one group's understandings and experiences dominate how knowledge is constituted so that there are no shared ways of knowing, and members of another group have no way to articulate their experiences to be collectively understood.[60]

The epistemic injustice Coyeen and Mundi experienced is intimately bound up with race, and colonial settlers' racism toward Aboriginal people, as Moreton-Robinson elaborates: 'The equation of whiteness with humanity secures a position of power from which whiteness reproduces itself and contributes to mainstream epistemologies' refusal of the specificity of the knowing subject'.[61] In my reading of Davis's interrogations of Coyeen, Mundi and Billy, understandings of the meanings of 'black' and 'white' shift, possibly because skin tone and culture are not the same thing, and/or possibly a problem of translation. The understood meaning of 'native' seems less inconsistent. Davis's statement of his conversation with Billy highlights aspects of Aboriginal humour and European ways of knowing. When Davis asks Billy whether the girls are black, he laughs. Davis asks him what he is laughing at and he says 'at his asking the question'. When Davis asks again 'were they black?', Billy says 'no'.[62] I wonder whether Billy laughed at Davis's question because the colour of the girls' skin was plain to see, and it seemed a stupid question. It seems Billy regarded Coyeen and Mundi as natives of Fraser Island whatever colour their skin was.

As far as I know, no dingoes played a part in Coyeen and Mundi's story. But the girls' story is important in relation to dingoes, not just because the man who interrogated them for the authorities in Sydney once accidentally killed an Aboriginal woman's beloved dingo. The epistemic injustice Coyeen and Mundi experienced parallels current authorities' omission to acknowledge and respect the diversity and individuality of relationships between Aboriginal people and dingoes. As a contemporary Butchulla man puts it:

---

60   Code 2008; Bailey 2014
61   Moreton-Robinson 2004, 87
62   Skinner 1974, 10

> There are different interactions with dingoes and humans – they are diverse. There are dingoes who are semi tame or in captivity … They are taking them as aggressive and therefore killing off all dingoes.[63]

I try to delineate how nineteenth-century settler attitudes toward dingoes on K'gari persist in the current Fraser Island Dingo Conservation and Risk Management Strategy logically but the rationale of these connections is more like a smell. It pervades.

* * *

Like James Davis, London-born convict and serial absconder David Bracewell came back to British-settled land with Petrie and Russell in May 1842. He escaped from Moreton Bay five times. Patrick Logan, commandant of the penal colony from 1826 until he was killed in 1830 by Aboriginal people near Mount Beppo, was cruel to convicts, known for his excessive use of the lash, and the number of escapees is staggering. The Moreton Bay Work Detail book records sixty-seven separate 'break outs' involving 123 convicts in the twelve months after February 1828.[64]

Bracewell spent nearly a year on K'gari in 1839–40 and observed thousands of people gathering on the eastern beach for the Feast of the Tailor Fish in late winter.[65]In contrast to Davis, who was guarded, Bracewell was talkative, and is credited with convincing Davis to come back to Moreton Bay.[66] Henry Russell believed Bracewell when he told him he had helped rescue Eliza Fraser but omitted Bracewell's alleged sexual liaison with, or rape of, her from his 1888 memoir *The Genesis of Queensland*.[67] *Stirling Castle* survivor Robert Darge confirmed that escaped convicts were at large when he told Curtis he met two 'bush-rangers' (the term for escaped convicts) called 'Tallboy' and 'Tursi'

63   Carter et al. 2017, 201 (Male X quoted)
64   Evans and Walker 1977, 42
65   Evans and Walker 1977, 43; Miller 1998, 31
66   Anon. 1889, 14
67   Schaffer 1995, 24–5, 93

living with Aboriginal people in the bush north of Moreton Bay.[68] By some accounts Bracewell's Aboriginal name, Wandi, means 'talker'.[69]

But Archibald Meston claimed that Bracewell's name Wandye means 'wild' or 'dingo'.[70] Less than two years after his return to European life, Bracewell died when he was crushed by a tree while he was timber felling. His cause of death matches that of another liminal border-crosser 100 years later: after evading bullets, baits and stirrup irons, Frank Dalby Davison's fictional dingo–kelpie cross Dusty was killed by a branch falling from a tree that 'put out his light as quick as a wink'.[71] The possible reasons for the naming of Bracewell are as enigmatic as why the sailors' heads were like dingo tails. The Butchulla named a man in terms of dingoes; they organised their perceptions of Flinders' mariners in terms of dingoes. Dingoes aren't just dingoes; they are a way of knowing.

In 1905, when the population of Aboriginal people around Cooloola and K'gari had been decimated and the reservation at Bogimbah on K'gari disbanded, Archibald Meston submitted to the Queensland Legislative Assembly a report outlining the island's history and economic resources, including timber and fish, and, presciently, its tourist potential. He thought that the *very numerous* dingoes on the island could be 'exterminated' to make it the perfect place for the preservation of native fauna.[72] Subsequently more of Kgari's timber was logged. Men who came for kauri, hoop and cypress pine, blackbutt, tallowwood and fibrous-barked satinay talked about predatory dingoes. '[Y]ou had to watch your dogs because the dingoes used to eat them. They'd try to cut the dogs out of your camp at night'.[73] Dingoes lured domestic dogs into the bush:

> A single dingo will approach a domestic dog and encourage it to play, all the time leading it further and further into the scrub.

---

68   Curtis 1838, 228
69   Evans and Walker 1977, 43
70   Evans and Walker 1977, 93, n. 27. I have kept spelling variations as they appear in sources.
71   Davison 1983, 240
72   Meston 1905, 5, italics in original
73   Williams 2002, 164 (Sid Jarvis quoted)

Other dingoes waiting in the scrub will join in the game until their instincts tell them the time is ripe. Then they turn on the dog and kill it.[74]

In 1950 the first sand-mining leases were granted; mining ramped up in the sixties and continued until conservationists were successful in shutting it down in the mid seventies. Logging ceased in 1991. In 1992 K'gari received a World Heritage listing.

No one in 1836 was preoccupied by what sort of agency dingoes on K'gari might have. As far as some of the *Stirling Castle* survivors were concerned, the dingoes had too much agency. They competed with the castaways for resources; they were allies, kin, of the Butchulla, who resisted and threatened the British invasion and settlement of their country.[75] But now K'gari is a World Heritage national park. It is meant to be looked after in a way that conserves the qualities that make it a World Heritage area: its outstanding natural beauty and the ongoing significant geological and biological processes that take place there.

QPWS manage K'gari's dingoes to preserve them as a species that contributes to biodiversity within a World Heritage 'wilderness'. The agency of individual dingoes, that is, their capacity to act in ways that produce the results that dingoes themselves intend, is not facilitated as part of this management plan. So conflicts occur, dingoes are killed, conflicts recur. The management strategy is a struggle. At the root of the struggle is the human insistence that dingoes need to conform to human ideas about how they should behave, whether that be 'species-typical' behaviour and/or a predetermined level of independence from humans. This insistence does not 'respect [dingoes'] agency', as Sue Donaldson and Will Kymlicka express it, but constrains it.[76] If dingoes can't be dingoes in a World Heritage-listed conservation area, where can they be dingoes?

Olga Miller, a 'Keeper of Records'[77] for the Butchulla people, describes how in the past people lived by the ethos of 'what is good

---

74   Williams 2002, 165 (Charlie Sorrenson quoted)
75   Evans and Walker 1977, 45–58
76   Donaldson and Kymlicka 2016, 2
77   Miller 1998, 31

for the Land comes first'.[78] But then 'strangers found this place'. They loved it too but they used metal to 'gouge and dig'. Sand mining left scars that made the moon, or Guggerie as she is known in Butchulla, cry. The rain brought new life and Yarrillee, the south wind, blew in the seeds of plants that would re-clothe the sleeping island.[79] For me, Guggerie's tears and Yarrilee's seeds make me think about my own sadness about the killing of dingoes on K'gari and my hope for finding other ways to know and live with different people and different species. In 2014 the Federal Court of Australia recognised the Butchullas' native title rights over K'gari but relationships between the Butchulla and the island's dingoes are still uniformly prohibited. Against such control, some Butchulla people assert their and dingoes' rights to territory and agency.

> Dingoes are like human beings. They communicate through their thoughts. They can tell when you're scared of them or not, they can tell if you're going to hurt them. If you sat down with them they'll sniff around you but it's against Parks and Wildlife's policies … but everything on the island is governed by what Parks and Wildlife wants you to do and what you can't do … It's a controversial area is the island where the government wants to dictate what happens and what goes on there. There's nothing the PBC [Prescribed Body Corporate], anyone, any Aboriginal people can do. That's the way it's going to be.[80]

K'gari shimmers and changes colour, obfuscated by stories. One moment the island is an idyllic tourist destination, a place for reckless fun and 'zee call of zee wilder', as depicted in the recorded message I listened to as I waited on the phone to book my two-day tour. In another instant it is a dangerous place where people have 'incidents' with wildlife and need the protection of strong authority, as evoked in the Queensland government recorded message I heard when I first rang to try to organise an interview with a QPWS ranger. There are also moments when people, even newcomers, are open, not full of hubris

---

78 Miller 1994, 5
79 Williams 1982, 176 (Olga Miller quoted)
80 Carter et al. 2017, 202 (Male X quoted)

and their own expertise. Harry Youlden, after suffering in fierce storms and cutting rain, learnt to survive the damp winter nights by warming his sleeping place 'with sand scraped from under the fire, after the manner of the natives'.[81] In these moments I believe we can learn other ways of living with dingoes, too.

81   Williams 2002, 44 (Harry Youlden quoted)

15

In the realm of science

*the deeper ground for my scientific inability seems to me to be an
instinct ... that ... has led me to esteem freedom more highly than
anything else*
Franz Kafka, *Investigations of a Dog*, 1922 [2017]

All the people I talked with cared about K'gari's dingoes, often in a
frustrated way. This kind of care is not new; it is 'one of power's forms'.[1]
On K'gari, dingoes are officially valued because they attract tourists and
because they are a unique 'remnant' population who contribute to the
biological processes that have given the island World Heritage status.
QPWS rangers who implement the Fraser Island Dingo Conservation
and Risk Management Strategy (FIDCRMS) have the unenviable job of
trying to keep people safe and trying to preserve a sustainable dingo
population, which, unfortunately, is not the same thing as protecting
dingoes – because individual dingoes are not considered important for
the conservation of the population as a whole.

Ostensibly, current dingo management on K'gari is driven by the
fear of another fatality. As far as I could tell, people who worked on
the island genuinely believed that killing some dingoes makes people
safe. The belief that humans have a right to safety wherever they are has

---

1   Probyn-Rapsey 2016, 41

become non-negotiable. It is part of a world view that, as legal scholar Ugo Mattei puts it, 'demand[s] that nature submit to human laws'.[2] Logging, mining – K'gari is no stranger to 'unsustainable extractive capitalism'.[3] Now tourism. Visitor numbers increase. Towns are fenced. Dingoes are excluded from the places they used to inhabit. They are vilified for acting like dingoes. Human words create dingoes who 'loiter', 'solicit' and 'steal'. More words are expended to justify the persecution and killing of dingoes; to argue that killing dingoes in the name of human safety on K'gari is necessary and not harmful to dingoes themselves. In this context, caring about K'gari's dingoes involves attempts to control people's perceptions of them and to control dingoes' behaviour.

Killing dingoes does not make visitors to K'gari safe. People hurt themselves diving into lakes and rolling their four-wheel drives. Killing dingoes does not result in a decline in the number of serious dingo–human incidents reported.[4] Some argue that killing dingoes increases the overall social instability of dingo packs and diminishes their territorial integrity, leading to heightened stress, elevated breeding rates, fatal dispersal of poorly socialised juveniles into neighbouring packs' territories and increased conflict between dingoes, and between dingoes and humans.[5]

QPWS staff emphasise their commitment to facts, not baseless opinions. According to the QPWS website, the current FIDCRMS:

has been prepared with expert input and is implemented by a team guided by qualified scientists who are wildlife experts in their own right. Collectively, the team has the most direct and consistent experience in managing dingoes on Fraser Island based upon long-term knowledge and understanding.[6]

---

2   Mattei 2016
3   Mattei 2016
4   Allen et al. 2015, 14; Ecosure 2012, 83, fig. 22; O'Neill et al. 2016
5   O'Neill et al. 2016
6   Ecosure 2013, Queensland government 2014

After Clinton Gage was killed QPWS ranger Terry Harper outlined the alternatives available to the island's managers:

> The Queensland government had three broad options available to it: a future where there were no dingoes, a future where there were no people, or a future where there are people and dingoes living together under a scientifically based management plan. That was the option that government chose.[7]

Dingo scientist Peter Fleming explains the principles of the scientific method: 'The obligation of all scientists is to appraise evidence (i.e. the data) and draw conclusions, develop hypotheses, test them and, when the evidence is strong enough, identify theories.'[8] But if the scientific method of dispassionately gathering and objectively appraising evidence, drawing logical conclusions, developing hypotheses and testing them with impartiality were practised on K'gari, what would the FIDCRMS look like?

Across much of Australia the killing of dingoes is encouraged by law, and government agencies and policies. Despite having lived here for at least several millennia, dingoes are considered an invasive species: pests. The animal-ethics application I had to fill out at my university before I could observe dingoes classifies them as 'exotic feral mammals'. Reports by state governments[9] and bodies such as WoolProducers Australia,[10] the Invasive Animals Cooperative Research Council[11] and the Australian Bureau of Agricultural and Resource Economics and Sciences[12] outline ways to manage and control – euphemisms for eradicate – wild dogs and dingoes. They are killed on pastoral land because they prey on livestock and in national parks under programs aimed at dingoes, foxes and wild cats. They are trapped and left to die, or trapped and shot, or baited with poisons such as, historically,

7   *Australian Story* 2011 (Terry Harper quoted)
8   Fleming and Ballard 2013, comments
9   NSW DPI 2012; Victoria DEPI 2013
10   WoolProducers 2014
11   Allen, B. 2011
12   Wicks et al. 2014

strychnine, and now 1080 (sodium fluoroacetate) or the newly available PAPP (4-aminopropiophenone).

Each state has different laws and, within states, these laws construe dingoes as native animals to be protected or pests to be killed, depending on where dingoes find themselves.[13] In Queensland, dingoes are classified as a Class 2 pest and land managers across most of the state are obliged to eradicate them. Queenslanders are not permitted to keep them so few people know individual dingoes independently of official government information. So, killing dingoes, even in a national park where they are meant to be protected, is routine, legal and acceptable to many people. Not surprisingly in a country where attempts to exterminate dingoes have been sustained for more than 200 years, implicitly their lives do not matter. Whatever efforts authorities have made to find the perpetrators, no one has been prosecuted for running down and killing one of the 2015 Eurong pups on 25 March 2016 or for deliberately poisoning six dingoes at Orchid Beach in June 2016.

The attempted eradication of dingoes has had profound impacts on dingo and human societies, on Australia's ecosystems and on scientific knowledge about dingoes. Until very recently, dingo research was dominated by agricultural interests.

Dingo authority Laurie Corbett, architect of the original Fraser Island Dingo Management Strategy and consultant to QPWS since the 1990s, worked with Alan Newsome at the CSIRO Division of Wildlife Research in the 1970s. Newsome's ten-year dingo project, with a focus on central Australia, was motivated by pastoralists' concerns about dingoes preying on domestic stock and was funded by the Australian Meat Research Committee. His research looked at dingo skull morphology (or shape), hybridisation between dingoes and domestic dogs, and predator–prey interactions. The team measured dingo skulls that were boiled clean of their flesh in a tin drum on a stony plain on the Barkly Tableland[14] and tested the efficacy of killing dingoes with 1080 and strychnine in various forms.[15]

---

13   WoolProducers 2014, appendices A and B
14   Newsome 2014, 163
15   Best et al. 1974

Newsome testified at the first inquest into Azaria Chamberlain's disappearance in 1980. Although both Newsome and Corbett did not think it was probable that a dingo had taken Azaria, they thought that the Uluru dingoes should be shot and that a 'fresh generation' of dingoes should be brought there and kept 'wild', because some of the dingoes at Uluru in 1980 were neither tame nor wild.[16]

Some of Corbett's research reflects an interest in sexual competition, infanticide and genetics – perhaps not unusual for a scientist of his era. His suggestion that dingo attacks on calves at waterholes on the Barkly Tableland in the Northern Territory were not motivated by hunger (often the calves were not consumed) but could have been 'displacement activity due to frustration over competition between males for access to breeding females, and resultant inter-male aggression' was cited as recently as 2015 in a study of conflict between people and dingoes on K'gari.[17] Corbett's experiment in which a captive alpha female ate the pups of the other females kept in the same forty-five- by twenty-five-metre enclosure for three breeding seasons from 1973 to 1975[18] cast a long shadow over female dingoes. No one I spoke with had witnessed this behaviour on K'gari. Parkhurst and Behrendorff told me about separate instances in which male and female dingoes had worked cooperatively to raise pups who were not their own offspring.

Corbett's view that dingo purity is threatened by hybridisation between dingoes and domestic dogs has been hugely influential,[19] enabling a 'violent logic of elimination'[20] against the 'impure' progeny of canids' illicit liaisons. In this context, so-called care for dingoes as a species becomes laden with the necessity to kill canids, called wild dogs, who are perceived not to be dingoes. Although anecdotally many people distinguish between dingoes and 'wild dogs', it is difficult to establish a consistent distinction because no description or collection of original specimens exists that defines the characteristics of 'pure' dingoes against which the level of hybridisation can be

---

16 Bryson 1986, 230
17 Appleby 2015, 140
18 Corbett 1988; 1995, 81–2
19 Corbett 2001, 2004
20 Probyn-Rapsey 2015, 57

assessed[21] and there is no diagnostic tool that can readily separate the lineages of dingoes and hybridised wild dogs to measure the genetic transfer between the two populations.[22]

The work of defining the dingo is ongoing, and not all experts agree. Wildlife ecologist Brad Purcell argues that dingoes should be defined by how they live and their role in the environment: they are hypercarnivores who eat more than seventy per cent meat; they live in groups; like wolves and unlike domestic dogs, they breed only once a year; they have variations in their genes and observable characteristics and traits.[23] Furthermore, their function shapes their form. Recent analysis of the cranial morphology, or skull shape, of dingoes, dingo–dog hybrids, 'pure-breed' dogs and cross-breed dogs shows that, in the offspring of dingoes and domestic dogs, dingo morphology is dominant.[24] There is morphological, or structural, overlap between dingoes and domestic dogs but dingoes tend to have a broader cranium than domestic dogs, a relatively larger palatal width, a relatively longer rostrum (snout), a relatively shorter skull height and a relatively wider top ridge of skull.[25] Hybridisation, and repeated interbreeding between dingo-hybrids and dingoes, lead to a canid with a dingo-shaped head. The authors of the study that compared the cranial morphology of dingoes and dingo–dog hybrids suggest that the shape of dingoes' skulls gives them a predatory advantage and enables them to hunt and survive more efficiently in the wild. Canids without dingo-shaped heads would be less likely to survive and reproduce.[26]

Back in the 1990s, one component of Corbett's plan to save dingoes from extinction via hybridisation involved setting up island refuges for 'pure' dingoes.[27] In 1998 he made the following recommendations regarding K'gari:

---

21 Crowther et al. 2014
22 Leonard et al. 2014
23 Purcell 2010, 40
24 Parr et al. 2016, 180
25 Crowther et al. 2014, 5
26 Parr et al. 2016, 172, 183
27 Corbett 1995, 177–8

assess and improve the purity of the dingo population on Fraser Island (by identifying and culling hybrids). Given that the extinction of pure dingoes on the Australian mainland is considered by some experts to be an eventual certainty, the preservation of dingoes on Fraser Island represents a unique opportunity to conserve the species in a natural environment. Fraser Island could become a model for the conservation of dingoes on other islands as well as confirming and enhancing its reputation of housing the purest dingoes in the world. This is likely to attract more tourists and enhance the dollar value of dingoes for commercial interests with the likely spin-off to the Department that external money could be obtained for continuing research and management of dingoes.[28]

The fear of interbreeding that underlies Corbett's proposed island refuges has parallels with colonial notions of the dangers of miscegenation between Aboriginal people and settlers.[29] Efforts to eradicate dingoes in one place and keep them alive in concentration camp-like conditions in another strike me as exercises in human hubris. It dawns on me that attempts by authorities on K'gari to avoid the aggression, which 'will rarely occur with wild animals that retain their natural wariness and distance from people',[30] by re-instilling in dingoes their allegedly 'natural fear of humans'[31] are part of a decades-long human-centred experiment in a bizarre form of 're-wilding'.

The current FIDCRMS was implemented after an extensive review of the 2006 strategy, which replaced the 2001 strategy. Critics of QPWS's dingo management had wanted the 2012 review to bring about change. The 2012 review is a 247-page document that goes into great detail about stakeholder surveys, interviews, questionnaires, governance, objectives and strategies.[32] There are graphs; there are case studies; there is multi-criteria decision analysis; there are risk

---

28 Corbett 1998, 17
29 Probyn-Rapsey 2015, 57
30 Ecosure 2012, 78
31 Ecosure 2013, 7
32 Ecosure 2012

assessments; and audits and reviews of governance models and terms of reference. Nevertheless, the review found that the objectives and strategies of the 2006 strategy were 'largely appropriate' and 'decision-making, methods of euthanasia and staff training are appropriate'.[33] Subsequently, the 2013 FIDCRMS found no need for change.[34] Incident reporting and killing dingoes would continue in the same way as they were already being conducted: 'decision making, methods of euthanasia and staff training (including the correct labelling and reporting of incidents) are appropriate'; 'the current level of euthanasia would be highly unlikely to impact on sustaining a viable wild dingo population'.[35]

As far as the information in the 2012 review goes, it was authored by an organisation called Ecosure, which is how I have referenced it.[36] No individuals' names appear on the publicly available copy I accessed. But according to a reference in another article about K'gari's dingoes,[37] the review was authored by B.L. Allen, J. Boswell and K. Higginbottom. Wildlife ecologist Ben Allen has published papers that argue that killing dingoes does not hurt native wildlife,[38] and that lethal control does not affect dingo home-range size or location.[39] Some of his research is supported by the Invasive Animals Cooperative Research Centre, an organisation – funded by the Australian government's Department of Industry and Science – that was set up to kill 'invasive' animals (a category that includes dingoes). Another of his publications is the *Glovebox Guide for Managing Wild Dogs: PestSmart Toolkit*.[40] 'Managing' here means killing.

Recently researchers have started to explicate dingoes' ecological roles. They argue that, as an apex or top-order predator (that is, the top of the food chain), dingoes drive trophic cascades by suppressing the abundance or altering the behaviour of meso-predators such as cats

33 Ecosure 2012, 3
34 Ecosure 2013, 5
35 Ecosure 2013, 5
36 Ecosure 2012
37 Allen et al. 2015
38 Allen et al. 2014b
39 Allen et al. 2014a
40 Allen 2011

and foxes.[41] Some scientists consider them to be a key to slowing the accelerating rate of extinctions that have been happening in Australia.[42] One study advocates the restoration of dingoes on cattle enterprises because they help improve biomass for grazing stock by predating on native herbivores.[43] In 2015, zoologist Lee Allen, Ben Allen's father, published an article that found that baiting changed the age structure and group size of dingo and wild dog populations, making them less efficient at hunting difficult-to-catch wild prey and more prone to predating on calves.[44] But, as is always the case with dingoes, not everyone agrees. Ben Allen maintains that not all claims made about the ecological roles of top predators can be substantiated by current evidence and that presently there is unreliable and inconclusive evidence for dingoes' role as a biodiversity regulator. He is a proponent of lethal control of dingoes.[45]

In 2016 the Allens became notorious for their plan to eradicate 300 goats on Pelorus Island in the Great Barrier Reef by releasing dingoes onto the island to kill and eat them. Two male dingoes were trapped on cattle and cane properties around Ingham, desexed and implanted with time-activated capsules of 1080 poison before they were released on Pelorus. In Ben Allen's photograph a scrawny black and tan dingo, who has been fitted with a radio-tracking collar almost as thick as his neck and has hip bones protruding from his rump, looks very nervous as he runs off after his release.[46] The 1080 capsules were allegedly a back-up that would kill the dingoes in two years if they hadn't been shot already. This precaution was taken because in 1993 Lee Allen was involved in a conservation program in which sixteen dingoes were used to eradicate 3000 goats on Townshend Island in the Defence Department's Shoalwater Bay military training area. The dingoes killed all the goats in two years, according to Lee Allen, but it was another ten to fifteen years before the dingoes were eradicated.[47]

---

41 Johnson et al. 2007; Letnic et al. 2012; Ripple et al. 2014; Ritchie et al. 2013; O'Neill 2002; Wallach 2011
42 Johnson 2006
43 Prowse et al. 2015
44 Allen 2015
45 Allen et al. 2013
46 Schwartz 2016a
47 Schwartz 2016a

The Allens looked forward to carrying out similar projects on many other islands[48] and seeing their conservation model adopted globally.[49] So they may have been surprised by the massive public outcry about the cruelty of their scheme. Queensland Environment Minister Steven Miles intervened and ordered Hinchinbrook Shire Council to remove the dingoes because, he said, they posed a threat to a vulnerable population of beach stone-curlews.[50] Despite the publicity surrounding the shutting down of the program, Hinchinbrook Shire Council were unable to catch the two dingoes and they remained on Pelorus. In September 2017 the mayor of Hinchinbrook Shire, Raymon Jayo, said that the plan to use dingoes to hunt goats was working well, that the island was starting to rejuvenate, that Ben Allen was evaluating the trial, and that they expected to go back to the government with their results in early 2018.[51] In mid 2018 the Queensland government was unforthcoming to inquiries from members of the public about what had happened to the dingoes.[52] In June 2019 the *Courier Mail* reported that one of the dingoes and one billy goat were still alive.[53] At a symposium in September 2019 Ben Allen explained how the dingoes 'removed' the kid goats rapidly, then the females, then the males, and that they had taken the goats to 'near nil'. He showed pictures of vegetation growing on Pelorus and claimed that the slopes where the goats used to live 'look fantastic now' and that the project had been a 'successful island recovery program', cheap and quick.[54]

Before I started researching dingoes I didn't know that killing in the name of conservation, cruelty to dingoes and to goats – who had had no experience of this kind of predation since humans released them on Pelorus Island in the 1880s as food for lighthouse keepers and shipwrecked mariners – and lack of government accountability constitute science. In conservation circles, I have discovered, it is radical to propose that 'Killing for conservation often proves to be

48 Schwartz 2016a
49 Vogler 2016
50 Schwartz 2016b
51 Cripps 2017
52 Parkhurst 2018
53 Michael 2019
54 Allen 2019

unjustified because although the costs to those individuals killed are certain, the benefits to populations and ecosystems are not'.[55] Scientists who advocate for compassionate conservation have adopted the principle: first do no harm. They look beyond the instrumental value of apex predators' role in conserving biodiversity and argue that 'Humanity has a moral obligation to help restore threatened populations, but harming sentient beings is a serious matter that cannot be justified solely on the basis of noble aims'.[56]

Ben Allen is the lead author of an article called 'Balancing dingo conservation with human safety on Fraser Island: the numerical and demographic effects of humane destruction of dingoes', which argues that killing what I calculate is an average of seven to ten per cent of K'gari's dingo population every year 'is highly unlikely to have adverse effects on dingo population growth rates or breeding success'.[57] The article argues that because QPWS usually kills juvenile male dingoes, who are more likely to be 'problem' dingoes, the breeding females and the overall population are not affected. When I first read the article, I wanted to ask the authors what would happen in their families if all the inquisitive, bold young males were killed. Nevertheless the article offers insights into how QPWS treats dingoes on K'gari. It provides details of numbers, locations, ages, genders and dates of QPWS's dingo killing. I learnt that May is the month of the greatest number of dingo destructions and that mostly juvenile male dingoes 'are removed' from places heavily frequented by people, such as Hook Point, Waddy Point and Eurong.[58]

According to the article, between seventy-six and 171 dingoes live on K'gari in nineteen family groups. If this estimate is correct, the number of dingoes is about the same as the number of humans who live on the island permanently. A graph entitled 'Population abundance estimates of dingoes on Fraser Island since 1992'[59] shows ranges from a low of seventy-six to a high of 239. What alarms me is that the

---

55   Wallach et al. 2015, 1
56   Wallach et al. 2015, 1
57   Allen et al. 2015, 1
58   Allen et al. 2015, 8
59   Allen et al. 2015, 8, fig. 2

low estimates in three recent studies are declining: 109 in 2004;[60] eighty-nine in 2011;[61] seventy-six in 2015.[62]

One of the article's most startling findings appears framed as yet another defence of killing. Dingoes are killed, allegedly, because they pose 'genuine human safety risks'.[63] But this killing does not affect the fitness of the population by removing successful breeders who might become aggressive toward people, resulting in a more docile population, Allen et al. argue, because killing dingoes does not result in a decrease in serious incidents.[64] Other studies corroborate this astonishing admission.[65] So, although many interactions with dingoes are neutral or benign, the relatively rare serious dingo maulings have led to 'illogical responses'.[66]

Even more astonishing, Allen et al. continue:

> destroying dangerous dingoes expected to breed (and produce even more dangerous dingoes) is arguably a goal of humane destructions anyway ... and may also reflect a continuation of the domestication process of dingoes and other dogs.[67]

It's no new thing for humans to decide how animals should be and to select for that behaviour, but it is confusing – domesticating animals to make them 'wild'. Critics of the current management strategy are concerned that, without better population data, the elusive 'wild' dingoes the authorities want on K'gari may not exist at all and that there are not enough dingoes on the island to sustain a viable population, let alone withstand the regular killing of 'habituated' dingoes who are not 'wild' but may be 'aggressive'. The ethical debate about whether dingoes should be killed or not and the potential effects on dingoes' health and wellbeing of trapping and ear tagging them, Allen et al. write, is outside

---

60 Baker 2004
61 Appleby and Jones 2011
62 Allen et al. 2015
63 Allen et al. 2015, 1
64 Allen et al. 2015, 14
65 Ecosure 2012, 83, fig. 22; O'Neill et al. 2016, E, fig. 2
66 Lennox, R.J. et al. 2018, 283
67 Allen et al. 2015, 14

the scope of their article.[68] No alternatives to 'humane destruction' are discussed in 'Balancing' and no reasons for dingo resilience, especially in relation to the Eurong pack,[69] are given.

According to the FIDCRMS, dingoes attracted to people by food become habituated, which leads to interactions, which can lead to aggression. The 'habituation' that QPWS are trying to avoid is more accurately described as conditioning. A food-conditioned dingo associates humans or the smell of humans with anthropogenic food.[70] According to this thinking, dingoes' sociability with humans is motivated primarily by food. Even though individual QPWS rangers recognise nuance and complexity in dingoes' behaviour and the feedback loops of action and response that occur when humans and dingoes interact, under the current management strategy 'habituated' dingoes are seen through the lens of outdated behaviourist theories that 'saw the animal as no more than an automaton for whom understanding is limited to simple associations'.[71] Without addressing dingoes' preferences and agency, the FIDCRMS seems impossible to implement. Counter to the FIDCRMS's reductive attempts to 're-wild' them, perhaps dingoes like Bold are conducting their own research, maybe even trying to re-domesticate humans. As the interaction reports record and as ranger Dan Novak observed, Bold was 'testing', 'exploratory testing' and 'learning about us'. His interspecies sociality wasn't always diplomatic from a human point of view, but it was creative.

In a discussion of the fieldwork of primatologist Barbara Smuts and wolf researcher Farley Mowatt, Vinciane Despret turns the lens around to consider the questions animals might ask humans.[72] When Smuts began her fieldwork she tried to habituate the baboons in Tanzania's Gombe Stream National Park to her presence by pretending to be invisible. But, unlike Smuts, the baboons did not believe in the 'so-called scientific neutrality of being invisible'.[73] It was only after Smuts had started to walk, sit, hold her body and address the baboons like a baboon

---

68   Allen et al. 2015, 4
69   Allen et al. 2015, 14, 15
70   Appleby 2015, 142
71   Despret 2016, loc. 319
72   Despret 2016, loc. 473
73   Despret 2016, loc. 466

that they began to treat her as a social being and someone with whom they could communicate. When they gave Smuts 'evil looks', Smuts understood she must distance herself. She considered that she had made progress because the 'possibility of conflict and of its negotiation is the very condition of the relation'.[74] Similarly, controversial wolf researcher Farley Mowatt had to learn how to be a 'politely accountable' observer in the field. Like Smuts, at first Mowatt, living alone in his tent in the middle of the territory of a pack of wolves, tried to be as discreet as possible. The wolves acted as though he didn't exist until one night, with the help of a lot of tea, he claimed ownership of his camp, urinating on every tree, shrub and tuft of grass that the wolves had already marked. Mowatt was concerned when the wolves returned from their hunt and, initially, walked past his tent, until one detected his marking and stopped in surprise, turned around, sat down and stared at him with 'uncanny intensity'.[75] Mowatt was anxious and turned his back on the wolf to communicate that the wolf's staring was contravening 'the most elementary rules of manners'.[76] The wolf started to carefully leave his own marks on top of Mowatt's. Now that his territory was 'ratified', wolf and human participated in the ritual of scent marking, one behind the other, each on his own side of their boundary. With the precedents of the baboons who communicated with Smuts and the wolves who negotiated with Mowatt, I interpret Bold's actions as an insistence that people 'learn either to ask that what matters to [him] be taken into account or to respond to such a demand'.[77] But, in the circumstances, conflict ensued. Sadly, Bold's trust that the negotiation of possible conflict was part of the relationship was misplaced. Perhaps, one day, dingoes like Bold will not pay such a high price.

QPWS cast Bold as a problem dingo but, as Linda Behrendorff pointed out, he was also a sacrifice: 'Is it the sacrifice of the one or two that actually saves the rest of the population?' All the people I spoke with wanted to avoid another cull like the one after Clinton Gage was killed in 2001, though Allen et al.'s sanguinity about such

---

74   Despret 2016, loc. 497
75   Despret 2016, loc. 535
76   Despret 2016, loc. 535
77   Despret 2016, loc. 540

culls is chilling. They cannot imagine how culls of over thirty dingoes could 'constrain or reduce overall dingo population growth on Fraser Island' if they are not conducted every year and there is no other additional cause of mortality such as disease.[78] So, I hypothesise, Bold and other dingoes killed on K'gari for high-risk behaviour are sacrifices in a religious, symbolic sense. They carry the sins and wrongdoings of the people who feed them or allow them to come too close, the dingoes who mauled and killed Clinton Gage, the dingoes who took and ate Azaria Chamberlain, and future dingoes who will approach people. They are killed to maintain order. Some people might think it is far-fetched to suggest that the idea of religious sacrifice plays a prominent part in a so-called scientifically grounded management plan. But dingoes are a submerged element in the settler Australian psyche. Systemic and deeply held subconscious beliefs that killing other predators is an act of care, and fears of miscegenation, are implicit in the work of prominent dingo scientists. One of the many interesting things about relations between dingoes and settler Australians on K'gari is that a move has been made from massacre or extermination, which happens on the mainland, to control and sacrifice on the island.

In a context where lethal control is commonplace and the welfare of individual dingoes does not matter, some researchers argue that non-lethal methods of controlling dingoes' behaviour are more humane.[79] These non-lethal, aversive conditioning methods include shock collars, aerosol sprays, non-toxic bitter baits or emetics (baits that make the dingo vomit), flashing lights, alarms and ultrasonic devices. Another method used on K'gari is hazing, or shooting dingoes with clay balls from slingshots, which taught dingoes to avoid the rangers delivering the aversive stimuli or their vehicles rather than to avoid people in general and the areas people frequented.[80] When Linda Behrendorff was nipped in 2009 she was helping animal behaviourist Rob Appleby to put a collar on the brother of the dingo who nipped her. Appleby conducted a pilot study on aversive conditioning by putting shock collars, also known as e-collars or electronic training devices,

---

78   Allen et al. 2015, 14
79   Appleby 2015, 154
80   Smith and Appleby 2018, 6

on dingoes who had been exhibiting behaviour that was a concern to rangers. Each of the four trials involved a different dingo. Responses of two of the dingoes suggest that shocks immediately stopped them from making physical contact with a person and approaching an unattended child. Another dingo became hesitant about approaching a backpack or a vehicle after a small number of shocks. The fourth dingo 'consistently fled after receiving a shock no matter what target behaviour was involved' according to one publication.[81] The information I have found about how this study was conducted and its outcomes is scant: single paragraphs in two research articles[82] and two devastating sentences in the 2012 review of the dingo management strategy, stating 'all dingoes fitted with collars were ultimately destroyed for subsequent incidents'.[83]

Why do I find the use of this kind of aversive conditioning on wild animals so repugnant? I have no qualms about using a deep growling voice and the word 'no' to try to stop my dog from eating rubbish in the park, or to squirting the cats from next door with water to dissuade them from spraying on our barbecue. But these are domestic animals. They get the good as well as the bad: the food, the shelter, the company, as well as the boundaries. When humans try to control wild animals with aversive conditioning, there's no kindness, no good part; it's all bad. And if my dog is anything to go by, the sensitive, smart dingoes of K'gari would be much more receptive to positive reinforcement – however that might be achieved. I'm not the only one with deep misgivings about the use of shock collars on canids. In its position statement on the use of electronic training devices, the European Society of Veterinary Clinical Ethology outlines shock collars' well-documented risks to dogs' health, behaviour and welfare, and calls for them to be banned.[84] Smith and Appleby, the authors of one of the studies that investigates aversive conditioning, contend that an in-depth review of the ethical and animal-welfare implications is 'beyond the scope' of their paper which is, admittedly, focused on using 'innovative methods' rather than killing dingoes to protect

---

81   Appleby 2015, 153
82   Ibid; Smith and Appleby 2018, 7
83   Ecosure 2012, 90
84   ESVCE 2017

livestock.[85] In response to criticisms from veterinarian Ian Gunn that his aversive-conditioning study was 'ethically questionable',[86] Appleby defended his research: 'The real dilemma is not whether aversive conditioning is ethically questionable ... but whether aversive conditioning is more ethically questionable than lethal control'.[87] If Appleby's argument here is that – because humans subject dingoes to lethal control, or kill them – aversive conditioning (which can be called torture) is more ethically acceptable, I do not agree. Murder does not make torture okay.

* * *

Despite what appears to me to be the tremendous callousness of some dingo science, empirical observation offers fascinating insights into dingoes' lives. An article published in 2013 tells the story of a dingo family's reaction when one of their members dies.[88] The observations take place at a dingo rendezvous site in open sclerophyll forest near Eurong. After pups emerge from the natal den when they are a few weeks old, they play and sleep and pass the time in soft places, welcoming clearings, or spots with other advantages such as cover or good sight lines.

Just before 5 p.m. on 9 September 2008, Appleby followed a three-year-old dingo mother and her four three-month-old pups to a regular rendezvous site. There he saw a fifth pup from the litter lying in a prone position in distress and convulsing. At a distance of fifteen metres from the sick pup he observed the dingoes for around two hours and filmed them.[89] A few minutes after Appleby first saw the prone pup, another pup came into view, snuff barked and exhibited play behaviour. Snuffing or snuff barking is a rapid exhalation of air through the nose and dingoes use it to communicate, especially when humans enter their territory. Or, likely, instances of snuff barking are

---

85   Smith and Appleby 2018, 10, 1
86   Gunn 2011
87   Appleby 2015, 154
88   Appleby et al. 2013
89   Appleby et al. 2013, 45

recorded by humans who have entered dingo territory. It seems to me to be some kind of warning. A few minutes later the pups' mother came into view, sniffed the prone pup's ano-genital region and briefly sat near the pup. She whined and her hackles appeared to be raised. The other pups came into view. One made contact with the prone pup, started to follow its mother, turned back to sniff the prone pup, then followed its mother. The mother disappeared from view with two pups following her. One of the remaining pups put its paw on the prone pup's shoulder. A few seconds later the mother came back. Three pups moved toward her, they all approached the prone pup, then the mother went to the top of the hill twenty metres away from the prone pup and all four pups returned to their sick sibling. They sniffed its mouth and its ano-genital area. They wagged their tails. Occasionally, the prone pup moved erratically, lifted its head, whimpered. A couple of minutes later the mother returned, sniffed the prone pup's ano-genital area and whined. Three of her pups approached her and she moved out of view. Two of the pups play-fought. One of them snuff barked and approached Appleby to within six metres, followed by another who picked up a stick in its mouth.

The observations continue. There's no climax. No plot that I can discern. I can't differentiate what is important, so everything is. The active pups and their mother were clearly conscious of Appleby's presence. The pups approached Appleby more than once and snuff barked; their mother went to the prone pup and sniffed its ano-genital area; she sat about a metre away from her sick pup, then went to rest about five metres away. The other pups played rough-and-tumble games. They chased one another, growled, yelped. The pups, whose playing was vigorous and would have involved nipping, did not bite the prone pup; their play behaviour was subdued – Appleby et al. do not go so far as to say gentle – around their sick sibling.[90] Mother and pups disappeared from view and then they all came back to the prone pup. The mother sniffed the prone pup, then again sat one metre away, and then rested five metres away. Two of the pups play-fought; one chewed on its mother's ear; the fourth pup came within two metres of Appleby, snuff barked and ran away. Another pup approached Appleby before it, too, ran away. The mother

90   Appleby et al. 2013, 43

moved three metres away, went to the prone pup, approached Appleby, lay down, rolled over, got up and disappeared. She came back into view close to Appleby, snuff barked and mouthed a stick. Then she disappeared into the low light in the direction of the prone pup. Appleby heard her whining and later saw her walking about five metres away from the prone pup with an unidentified object in her mouth. He heard a pup yelp some distance away.

Before Appleby left he went to the prone pup and recorded, via night shot, that its mouth was open and its right eye appeared to be swollen. It showed no visible signs of life. As he started to leave, the pups' mother returned, approached, walked past him and sat one metre away from the prone pup. She briefly self-groomed, got up, walked away from Appleby and rested about five metres away from him before she moved out of view. Appleby waited a few more minutes before he left. Was she trying to communicate something? What's the significance of her and her pups' actions? Narrative demands a plot, some form of suspense, some reason for action. Otherwise, why include it? What might make sense to dingoes appears to humans as random, repetitive, aimless actions occupying moments better spent doing something else, like conducting an experiment or creating a narrative.

Appleby presumed that the prone pup died after he had been observing the dingo family for about twenty to thirty-five minutes, at around 5.15 to 5.35 p.m., because he could no longer see its abdomen rising and falling with breath, and the pup made no visible responses to its mother's and littermates' 'tactile investigatory behaviour'.[91] He didn't know what killed the pup but he assumed it was snakebite.[92] A table, 'A selective summary of the events observed during and immediately following the death of a pup', combines Appleby's 'ad libitum sampling protocol' observations[93] with data from the two digital video cameras he used to film the scene:

91   Appleby et al. 2013, 43
92   Appleby et al. 2013, 43
93   Appleby et al. 2013, 43

17:22:00 Pup approaches D-pup [prone pup] and exhibits oral and shoulder/chest/front leg sniffing then moves away. D-pup shows no response

17:27:44 Pup approaches D-pup and exhibits ano-genital sniffing then moves away. D-pup shows no response

17:29:00 Pup approaches D-pup, possibly exhibits a single, muffled snuff bark then moves away. D-pup shows no response. A second pup approaches, briefly sniffs D-pup, then both pups move away[94]

I cannot tell from these observations whether the same pup returned to sniff the sick pup or whether each of the prone pup's siblings approached one by one at intervals to check on it. A few photographs accompany the article and one shows four pups around the prostrate body of the fifth pup.[95] One pup stands and sniffs the prone pup's mouth, another standing pup sniffs or licks its genitals, one pup sits and looks toward the camera and another sitting pup looks as though it is watching its sniffing siblings. In the foreground between the camera and the pups lies a log. It is parallel with the body of the prone pup, whose head is facing downhill. The ground behind the log on which the pups lie, stand and sit is covered with leaves. Behind the furry forms of the young pups rise the twisted trunks and branches of eucalyptus fading into green foliage. The significant moment – the prone pup's last breath, last heartbeat – is quiet, unmarked; lived, like so many aspects of dingoes' lives, beyond the realm of human knowing. But I am sure the prone pup's mother and siblings knew something had changed. Dingoes, who communicate with breath – through inhalations, exhalations, scents, snuffs, snorts and sighs – would have known when the prone pup stopped breathing. Something changed everything. After a while those warmth-seeking, tactile dingoes would have known that the prone pup's body was no longer warm.

When Appleby returned the next morning, 10 September, the pup's body had been moved and was lying in a 'shallow digging consistent with a sleeping site'.[96] The next day, 11 September, Appleby found the

94 Appleby et al. 2013, 45
95 Appleby et al. 2013, 44, fig. 1A
96 Appleby et al. 2013, 44

pup's body had been moved again. He filmed the mother and pups as they walked past him to the dead pup. They all looked straight at the camera as they passed; the mother's hackles were up. At three months, the pups were beyond the stage when they would usually be carried by their mother, but Appleby filmed her putting her mouth around the abdomen of her pup, near its hind legs, to pick it up. Her grip must not have been strong enough or balanced properly because she dragged its body a little before she adjusted her grip, picked it up and ran away, without a backward glance at her trailing pups or Appleby, her tail curled over her back and tending slightly to her right. Two of her pups followed her; the last turned to look at Appleby from the spot where its dead sibling had lain. Once again, the dingoes showed through their gaze that they were conscious of Appleby. I wonder what effects his presence had on their behaviour. The mother walked about thirty metres south carrying her pup's body and laid it on the ground before she and the live pups rested about twenty metres away. Soon after, Appleby left the area and when he returned on 13 September he could not find the pup's body.

The authors comment on how the dingoes did not consume the pup's body, whose carcass was intact, during the observation period. They hypothesise that the mother may have been attempting to keep the dead pup close to her during periods of rest. They note the potential threat Appleby as observer may have posed to the dingoes. They discuss this dingo family's behaviour in the context of how orangutans, dolphins, elephants, giraffes and otters respond to a conspecific's death. They analyse a mother's attachment to the body of her dead young in terms of evolutionary biology – as adaptive or maladaptive behaviour – and try to assess 'whether, or to what degree, death is "understood" by those [conspecifics] responding'.[97] The word 'understood' is in inverted commas, denoting, I think, a certain discomfort about using it.

But I am interested in the thing that Appleby et al. do not name: grieving. The dingo mother's response to the change of state of her pup was to seek privacy and protection. This is mine, she said to Appleby as she passed with her hackles up before she grabbed her pup's body and ran away. Did she keep her pup's body close to her because the pup might wake up? Because of Appleby's constant presence? Or because she did not want

---

97   Appleby et al. 2013, 44

to leave the pup? I am not interested in other animals' thanatology or how a mother's response to the death of her offspring relates to evolutionary biology. I am myopic. I am interested in this pup, this pup's siblings, this pup's mother. I am interested in this pup's round little body in this clearing, on this leaf-strewn ground, in this dappled shade.

Coda

one does not die so easily as a nervous dog imagines
Franz Kafka, *Investigations of a Dog*, 1922 [1946]

The tide was about halfway in when a group of seventeen International Student Volunteers (ISV) and their tour leader went down to the beach at Eurong to see the sunrise on the morning of Sunday 16 August 2015. The students' two weeks of meaningful volunteer work during their time in Australia might have included conserving natural sanctuaries for wildlife and restoring and protecting habitat before they were taken on a two-week adventure tour that included K'gari and other parts of the east coast. When I accessed it, the ISV website promised volunteers the opportunity to see Australian wildlife in natural settings as they contributed to projects that aided the survival of these animals.[1]

Stars so clear the night after the new moon; the darkest nights, the brightest stars. After the sun rose at 6.18 a.m. but before the tiny sliver of crescent moon appeared just before seven, PuGY walked past the group of volunteers and approached two young men who were standing about fifty metres away from the others. They followed their tour leader's shouted instructions, stood up straight and faced the dingo with their arms folded. But PuGY darted behind one of them and bit

---

1   ISV 2018

him, leaving two scratch marks on his left thigh. He then approached the group, and sniffed and attempted to bite a camera on a tripod before he ran off northwards toward some fishermen about 250 metres away.

The students went back inside the fence. They were due to catch a barge to the mainland soon after so rangers did not speak with the nineteen-year-old who was nipped. The tour leader emailed photographs of the small marks PuGY left on his skin, one of which was included in the interaction report. The names of the reporting officers do not appear on QPWS's interaction report forms, just their numbers. The officer who reported this incident was the officer who gave PuGY his first interaction report, a Code C, on 28 March when the eight-month-old dingo woke him up by sniffing his head when he was sleeping in a swag at Cornwells camping zone.

Later that day, near Eurong, rangers administered the anaesthetic Zoletil 100 from a blowpipe to PuGY from close range. They didn't need to trap him because he approached people so recklessly. Zoletil 100 contains 250 milligrams of tiletamine, which inhibits the action of the N-Methyl-D-aspartate receptor (a glutamate receptor thought to be important for synaptic plasticity, learning and memory) and is pharmacologically similar to ketamine though it is more potent and longer lasting, and 250 milligrams of zolazepam, a benzodiazepine, which is a tranquilliser that has sedative, hypnotic, anti-anxiety, anticonvulsant and muscle-relaxant effects, though paradoxical reactions such as aggression, disinhibition, worsened agitation or panic can occur. The analgesia and immobilisation provided by the tiletamine complement the muscle relaxation and tranquilisation provided by the zolazepam.[2]

According to its manufacturer, Zoletil's effects include fast catalepsy (which is a trance or seizure with a loss of sensation and consciousness accompanied by rigidity of the body) followed by muscle relaxation; and moderate superficial, immediate and visceral analgesia (which means the large organs – heart, stomach, lungs, intestines – do not feel pain). It does not affect important reflexes such as the laryngeal, the pharyngeal and the palpebral, and it does not depress the lower cranial nerves.[3]

---

2   Lee 2006b, 11
3   Virbac 2015

QPWS staff sedate a dingo with Zoletil only when they need to do invasive procedures, such as ear tagging or microchipping. For a couple of hours while their patient is under the effects of the drug, rangers tag, chip, check the dingo's health, and take measurements and a DNA sample, which might be a few hairs. Once these procedures are done the dingo might be sitting up, lucid, but rangers need to ensure the Zoletil has completely worn off before they release their patient. Often this processing of dingoes happens at night. Rangers might put the dingo, covered with a blanket, in a cage for a few hours in the field rather than taking it back to a ranger station. They let it go near the place they trapped it and, ideally, near water when it is completely recovered – alert and awake.

Zoletil has variable effects on canids. The onset of sedation can range from a few to up to ten minutes, and after sedation some canids are 'just capable' of walking and some are almost unconscious.[4] Unlike general anaesthetics, dissociative anaesthetics do not necessarily involve a loss of consciousness. They disassociate the thalamocortic and limbic systems[5] and interrupt neuronal transmission from unconscious to conscious parts of the brain.[6] The patient does not appear to be anaesthetised and can swallow and open their eyes but does not process information. Emergence from Zoletil may be accompanied by delirium, excitement, disorientation and confusion. Canids may be restless when they recover.[7] Dogs are described as having 'slow and stormy' recoveries accompanied by 'head shaking, salivating, muscle rigidity, vocalization, defecation'.[8]

How PuGY had recovered from Zoletil in the past, when he was tagged in March, when he was fitted with the radio-tracking collar in July, was of no consequence now. After he was sedated he was taken to the ranger base at Eurong, where his heart was injected with a clear green fluorescent liquid called Valabarb, pentobarbitone sodium. Pentobarbitone is a short-acting barbiturate, also called Nembutal. If the people administering the injection were following the

4   Clarke and Trim 2014, 435
5   Lee 2006a, 4
6   Lee 2006b, 10
7   Clarke and Trim 2014, 435
8   Lee 2006b, 10

manufacturer's safety instructions, they would have been wearing safety glasses with eye shields or chemical goggles, chemical-protective PVC gloves and rubber footwear. They might have looked and smelt a bit strange to PuGY but he would have been feeling strange anyway. If he wasn't already unconscious he would have been soon after the Valabarb was administered. He would have experienced central nervous system depression resulting in decreased heart rate, a decreased rate of breathing, and loss of consciousness. Within one or two minutes his heart and brain function would have shut down. Profound coma and death would have followed.[9]

His body was kept in a freezer, possibly at the ranger station at Eurong, until it was defrosted and a vet, observed by four rangers, performed a necropsy on 15 October. PuGY's two-page dingo necropsy checklist is both impersonal and intimate. As well as giving measurements of his head length (230 millimetres), ear length (140 millimetres), tail length from skin to tip (320 millimetres), body length (730 millimetres), neck width (385 millimetres) and weight (18.2 kilograms), it relates that his skull was not kept but his ear was in the freezer. He was recorded as being in good condition, rated four out of five, with good muscle mass, hips and ribs covered but noticeable.

His stomach was eighty per cent full of mostly fish (forty per cent); bait (squid and prawn) (twenty per cent); fur (mammal) (twenty per cent); leaf litter (vegetation) (fifteen per cent); and citrus (five per cent). His lungs were a normal colour and in good condition. His blood vessels, trachea and bronchus were clear. His pericardium, the tough, double-layered sac that protected his heart and the roots of his great blood vessels, was listed as normal, as were his liver and kidneys. His gall bladder was eighty per cent full. In the slurry in his large intestine the vet found hookworm and whip worm. His spleen was 'round and swollen consistent with barbiturate use'.

On the next page are photographs: a close-up of the round purple, green and yellow ear tag punched through the wide part of his left ear; the pads of his two front paws with their unique arrangement of light and dark tones, pink and dark grey; his lips held back by a left hand in a pale blue surgical glove reveal his excellent teeth, a wall of upper and

---

9   Jurox 2015

lower incisors terminating in the long, shearing prongs of his canine teeth; another, or the same, gloved hand holds his dark-furred tail with its tip of white hairs.

The next four photographs of his internals are coloured crimson and dark liver-brick red. It is hard to gauge the scale of the picture captioned 'Abdominal cavity', which shows PuGY's pelt sliced open to reveal layers of pink, dark red and crimson, also sliced open. In another photograph, a fan-like structure comprising red lattice around transparent areas is held by three gloved hands and captioned 'Abundant Omental fat stores'. In another his heart has been sliced open. It looks big with one half resting on one of his golden legs. The caption reads, 'Heart, normal, fat coverage – 40%, 3mm.' The last picture shows his shiny dark-red kidney grasped between the thumb and forefinger of a light blue plastic glove: 'Kidney, normal, fat coverage – 20%, 4mm.' Whatever purpose omental fat stores served for Bold, for me, writing about his heart and his kidney, his teeth, his tail and his paws, keeps him here. I am not abandoning him. I can somehow keep him alive.

According to QPWS's necropsy, PuGY was a healthy dingo. But the National Dingo Protection and Recovery Program sent his necropsy to some other vets, and they claimed that the photograph of his abdominal cavity showed massive internal bleeding 'consistent with a heavy blow or impact prior to being put down through lethal injection to the heart'.[10] Veterinarian Ian Gunn was critical of the inadequacies of the necropsy: 'We have evidence of unacknowledged animal trauma and unanswered animal welfare questions.'[11]

Euthanasia comes from the Greek words for good death. The Collins online dictionary defines it as 'the practice of killing someone who is very ill and will never get better in order to end their suffering, usually done at their request or with their consent'. So the killing of PuGY and others like him is not, strictly speaking, euthanasia. Even a death intended to be gentle and easy involves violence in the form of the force or energy needed to kill, to create a breakdown in the body, to make the heart or the brain cease, to make blood not go where it should or go where it should not, to cause an extinguishment.

10   NDPRP 2016, 43
11   NDPRP 2016, 43

Perhaps the sedative did not work well for Bold. Perhaps he became agitated. Perhaps one of the people charged with the bad job of killing him was frustrated with his never-ending antics. Perhaps he fought the drowsiness of the Zoletil. Perhaps he fought the people who took him to the ranger station to inject his unrepentant heart with Valarbarb.

Had PiYPi, 08Purple or 08Red investigated what happened, and I can't imagine curious dingoes would not investigate, they would have been able to sense, through smell, that when PuGY was taken into the building he was alive. To some extent they would have been able to work out what happened and the order in which it happened. Perhaps, when no one saw them, PiYPi or 08Red or 08Purple felt their bowels move and, voluntarily, involuntarily, defecated in the middle of the doorway through which Bold had been taken. Dingoes sometimes leave a scat at the door of the ranger station when one of their pack members is killed by Parks. They also would have been able to smell the parts of PuGY's body that came out of the freezer two months later.

The moon was a waxing crescent the night after Bold's necropsy. Like the new moon on the last night he was alive, it did not cast much light. If his body was buried inside the fence that encloses Eurong, PiYPi might not have followed his son's scent to the place where the sand was disturbed. If he did, he would have stayed out of sight, as required under the FIDCRMS, because he knew that, with humans, 'agreement is the best weapon of defense'.[12] Did he have to decide whether to track his son's head, which was disposed of separately to his body, or his body? Or did he sniff out both? Perhaps he dug, his head craning down, his ears forward, his tail curled over his back, cool sand flying behind him in a flurry and the smell of PuGY's body filling up his muzzle as he got closer and closer. Sometimes, after QPWS staff bury a dingo, other dingoes dig down, just deep enough to reach its body. They don't eat it. What would PiYPi have made of the incisions where Bold's heart and kidneys were removed, the way his ear was sliced off? That humans and dingoes are interested in the same things? He would have recognised the unique scent of his son's paws. Perhaps it confirmed something for him. Perhaps this practice of digging down to the body of a dingo killed by human hands is part of some dingo ceremony, PiYPi's big paws removing the

---

12   Kafka 1946, 33

weight of sand from Bold's body, freeing his raucous spirit, enabling his son to shine through him to the brilliant distant stars.

The 2015 pups, gangly, skinny, would have come to greet him when he got back to the den. They would have crouched before him, tails between their legs, each one licking his mouth, inhaling the smells of where he had been, what he had eaten. 08Purple would have greeted him too, nose to nose, their tails curled over in courteous symmetry, two halves of one heart.

After his brother was killed, 08Red moved north. Luckily for him there are only two mentions of him in the interaction reports from April to November 2016. In early June 2016 he received a Code C for passing a car and some campers north of Woralie Creek on the west coast, without threatening anyone or causing any issues. The next day he received another Code C for loitering around two fishermen on the beach at Connors Corner near Waddy Point on the east coast. His red ear tag was starting to fade to pink. In early 2018, 08Red was on the remote north-west coast, a long way from the tourists of Eurong. He was seen with a female dingo, but people did not know whether they had raised pups.

For his younger siblings, the 2015 Eurong pups, there were many more reports. One, a female, had a thing for teddy bears and pillows, which she took from campsites. With her brothers and on her own she approached tourists and children playing badminton on the beach at Eurong, and attempted to lick the foot of a child whose family were camping near One Tree Rocks. In late March 2016 one of her brothers was killed by a vehicle. On 3 April another brother was killed by Parks for high-risk behaviour. Another died in mid June from indeterminate causes after he had been fitted with a GPS collar. Perhaps the sister who was interested in teddy bears and badminton was the one who had stood between her mother 08Purple and our tour bus, her hips sticking out of her hide like crescent moons, an oversized head on her skinny body, her puppy muzzle dark and powerful, and her big eyes inquisitive and trusting. Year after year the Eurong young must learn how to live, as though 'it is the thousandth forgetting of a dream dreamt a thousand times and forgotten a thousand times; and who can damn us merely for forgetting for the thousandth time?'.[13]

13   Kafka 1946, 32

*　*　*

On a sweltering day in late January 2017 my sister's family and my children and I gathered at our mother's house. It was twenty-five years since my brother Bruce died of AIDS. We looked at photos of him walking my sister down the aisle on her wedding day. After Bruce died I often thought about what happens to the love you feel for a person when they're no longer alive. Where does it go? Now I wonder whether the same kind of question could be asked about the wat'dha. Where do they go? How can they just cease to exist? I told my mum and my sister that I thought about Bruce and missed him a lot, but he would have said, 'Enjoy the people you're with now.'

That afternoon the southerly came. The next morning I woke early, clear-headed and energetic. While my children played their e-devices I read Vinciane Despret on whether one 'can lead a rat to infanticide'. Her subject matter – field reports of infanticide among primates and the ensuing laboratory studies that created the conditions in which rats would kill their young – is dark and serious but her style is dry and funny:

> Rats, who until now had tested every drug possible, who were plied with alcohol and cocaine, who ran labyrinths for the behaviourists, inhaled thousands of cigarettes, knew experimental depression and neurosis, learned how to measure time ... these loyal servants of science had thus become infanticidal!
>
> They must be given credit: as will be seen, rats are not particularly adept at this kind of behaviour – but then again, they're not particularly happy about smoking cigarettes, testing drugs, or running labyrinths while hungry either. [...] Rightly or wrongly – they were not asked for their opinion – rats would become infanticidal.[14]

I was in the middle of redrafting the contact history of K'gari for the fifth or ninth time and I was pleased to feel so enlivened by this writing about animals and the people who study them. I thought Despret's

---

14　Despret 2016, loc. 1912–9

critique of how the 'sociobiological theory of intrasexual competition' explains all behaviour in terms of its adaptive value and ignores other possible motivations such as pleasure, or the social habits of the group, or the fact that relationships are important,[15] could be applied to some of the studies of K'gari's dingoes. It was a quarter to ten and I hadn't made my kids' breakfast when the telephone rang – one of my mother's neighbours. She said my mother had had a stroke. I should come.

Mum had been swimming in the sandy bay near her house in shallow water made choppy by the southerly, lighter now than yesterday. She didn't want to drown. She struggled and called out. She told the neighbour who helped her that she was having a stroke. When I got there she was lying on her side on a small stretcher on the wet sand near the water's edge. The tide was on the way out. Paramedics were attending to her. Her voice was strained, deep and gruff, but she was conscious and she knew what day it was and who I was. I can't remember if she was still holding her pink bathing cap and her left earplug. If not, they were nearby. I didn't find her right earplug.

She lost consciousness in the ambulance on the way to hospital. A neurologist at the hospital showed my sister and me images of the massive haemorrhage that had started in the right side of her cerebellum and gone 'everywhere', as he put it. The blood had gone into her brain stem. It was sending her blood pressure crazy and shutting down her respiration and her heart. I thought of a river eddying into the sea. Her breathing stopped before her heartbeat, which continued faintly for what seemed like a long time.

There will be no more stories like the one about my father driving in reverse down Urangan Jetty.

I kept writing. My obsession with dingoes changed a bit. It was still about death. But now it was about letting the dead rest, and letting the dead and the living go free.

* * *

My dog was sixteen the summer after my mother died. She came to the beach with us, a place we were all happy, and stood chest-deep in water

---

15   Despret 2016, loc. 1959–68

with waves washing over her, watching us in the deeper water, and bounding over to whoever the waves brought in to her in the shallow water. When I got out and walked up the beach she sprang, twisted, ran in big circles around me, barked and play-bowed.

When she sleeps – which is a lot – she is as delicate as a young pup. If I touch her she startles and springs to her feet, sprightly for an old dog. I don't like to wake her abruptly. So I lie beside her and say her name or make soft kissing noises. But she does not hear well now. In her dreams, I wonder, is she able-bodied? No arthritis in her back legs, her senses acute. I inhale the leathery, nutty, minty, black-tea scent of her paws. I put my arm near her nose so she can smell me and maybe smell herself awake. Awake or asleep, she prefers fresh air, her own air, her nose unencumbered by cloying people. But she does not wake. Perhaps the smell of my palm – sweaty, garlicky – will rouse her. She continues to sleep. I cup her breath in my hand.

Works cited

ABC news 2012. As it happened: Azaria Chamberlain inquest, 14 June, https://ab.co/3pZ4mbV.

ACA (A current affair) 2012. Tracy Grimshaw interview with Lindy Chamberlain, 12 June, https://bit.ly/2UZ6Mcr.

Alfredson, K. 2001. Australia culls dingoes amid public outcry, CNN.com./world, 3 May. https://cnn.it/2HAk7EX.

Allen, B.L. 2011. *Glovebox guide for managing wild dogs: PestSmart toolkit,* Invasive Animals Cooperative Research Centre, Canberra.

Allen, B.L. 2019. Ecological roles of dingoes: contributions from the Pelorus Island goat project, presentation at *The dingo dilemma: cull, contain or conserve,* Royal Zoological Society of NSW forum, Taronga Institute of Science and Learning, Sydney, 7 September.

Allen, B.L., P.J.S. Fleming, L.R. Allen, R.M. Engman, G. Ballard 2013. As clear as mud: a critical review of the ecological roles of Australian dingoes, *Biological conservation*, no. 159, 158–74.

Allen. B.L., R.M. Engeman, L. K.-P. Leung 2014a. The short-term effects of a routine poisoning campaign on the movements and detectability of a social top-predator, *Environmental science and pollution research international*, vol. 21, no. 3, 2178–90.

Allen, B.L., L.R. Allen, R.M. Engeman and L.K.-P. Leung 2014b. Sympatric prey responses to lethal top-predator control: predator manipulation experiments, *Frontiers in zoology*, vol. 11, no. 56, doi.org/10.1186/s12983-014-0056-y

Allen, B.L., K. Higginbottom, J.H. Bracks, N. Davies and G.S. Baxter 2015. Balancing dingo conservation with human safety on Fraser Island: the numerical and demographic effects of humane destruction of dingoes, *Australian journal of environmental management*, doi.org/10.1080/ 14486563.2014.999134.

Allen, L. 2015. Demographic and functional responses of wild dogs to poison baiting, *Ecological management and restoration*, vol. 16, no. 1, 58–66.

Allen, R. 1987. Man 'held up' by .38 towel, *Brisbane Sunday Mail*, 24 May.

Amperodirect 2017. Pulse assessment with pulse oximeter, https://bit.ly/3nW7qnj.

Anonymous 1889. Obituary of James Davis, *The Week*, 11 May, https://bit.ly/ 39bFGHe.

Anonymous 2012. Tears of despair for Fraser Island dingo, SFID media release, 24 November, https://bit.ly/2V6SZAh.

Appleby, R. 2015. Dingo–human conflict: attacks on humans, in B. Smith (ed.), *The dingo debate: origins, behaviour and conservation*, CSIRO Publishing, Clayton South VIC, 131–58.

Appleby, R. and D. Jones 2011. *Analysis of preliminary dingo capture-mark-recapture experiment on Fraser Island*, Final report to Queensland Parks and Wildlife Service, Griffith University, Nathan.

Appleby, R., B. Smith and D. Jones 2013. Observations of a free-ranging adult female dingo (*Canis dingo*) and littermates' responses to the death of a pup, *Behavioural processes*, no. 96, 42–6, https://bit.ly/3m62ifY.

Appleby, R., J. Mackie, B. Smith, L. Bernede and D. Jones 2017. Human–dingo interactions on Fraser Island: an analysis of serious incident reports, *Australian mammalogy*, https://bit.ly/2J8r3td.

Ardalan, A., M. Oskarsson, C. Natanaelsson, A.N. Wilton, A. Ahmadian and P. Savolainen 2012. Narrow genetic basis for the Australian dingo confirmed through analysis of paternal ancestry, *Genetica,* vol. 140, 65–73, doi:10.1007/ s10709-12-9658-5.

Armitage, E. 1923. Corroborees of the Aborigines of Great Sandy Island, written and translated by Edward Armitage, of Maryborough, Queensland; Vocabularies of four representative tribes of South Eastern Queensland in F.J. Watson (ed.) *Supplement to the journal of the Royal Geographical Society of Australia (Queensland)*, vol. 48, no. 34, 96–7.

Auster, P. 1999. *Timbuktu*, Faber and Faber, London.

Australian story 2011. Dogs of war, ABC TV, 28 February, transcript, https://ab.co/ 374TjoZ.

Bailey, A. 2014. The unlevel knowing field: an engagement with Dotson's third order epistemic oppression, *Social epistemology review and reply collective*, vol. 3, no. 10, 62–8.

Baker, N. 2004. *Preliminary report on the ecology of the dingo on Fraser Island*, University of Queensland, Gatton.

Baldwin, J. 1965. The white man's guilt, *Ebony special issue: the white problem in America*, August, 47–8.

Balme, J., S. O'Connor and S. Fallon 2018. New dates on dingo bones from Madura Cave provide oldest firm evidence for arrival of the species in Australia, *Scientific reports*, no. 8, article no. 9933, doi:10.1038/s41598-018-28324-x.

Basedow, H. 1925. *The Australian Aboriginal*, F.W. Preece and Sons, Adelaide.

Bates, D. 1985. *The native tribes of Western Australia*, ed. Isobel White, National Library of Australia, Canberra.

Behrendorff, L. 2015. Interview with Rowena Lennox, Eurong QLD, 26 November.

Behrendorff, L. 2017. A prickly subject: innovative handling of a difficult prey, *Australian mammalogy*, doi.org/10.1071/AM17024.

Behrendorff, L. 2018. Clever girl? An observation of innovative prey handling by a dingo (*Canis dingo*), *Pacific conservation biology*, doi: 10.1071/PC17044.

Behrendorff, L. and B.L. Allen 2016. From den to dust: longevity of three dingoes (*Canis lupus dingo*) on Fraser Island (K'gari), *Australian mammalogy*, no. 38, 256–60, doi: 10.1071/AM17024.

Behrendorff, L., L.K-P. Leung, A. McKinnon, J. Hanger, G. Belonje, J. Tapply, D. Jones and B.L. Allen 2016. Insects for breakfast and whales for dinner: the diet and body condition of dingoes on Fraser Island (K'gari), *Scientific reports*, no. 6, article no. 23469. doi: 10.1038/srep23469.

Behrendorff, L., G. Belonje, B.L. Allen 2017, Intraspecific killing behaviour of canids: how dingoes kill dingoes, *Ethology, ecology and evolution*, doi: 10.1080/03949370.2017.1316522

Bell, J. and A. Seed 1994, *A dictionary of the Gubbi-Gubbi and Butchulla languages*, no publisher, Brisbane. Digitised by Australian Institute of Aboriginal and Torres Strait Islander Studies, https://bit.ly/3q3GnbM.

Berndt, R.M. and Berndt, C.H. 1942, 'A preliminary report of field work in the Ooldea region, western South Australia (continued)', *Oceania*, vol. 13, no. 2, pp 143–69.

Berndt, R.M. and Berndt, C.H. 1977, *The world of the first Australians*, Ure Smith, Sydney

Best, L.W., Corbett, L.R., Stephens, D.R. and Newsome, A.E. 1974, 'Baiting trials for dingoes in central Australia, with poison "1080", encapsulated strychnine, and strychnine suspended in methyl cellulose', CSIRO Division of Wildlife Research technical papers, no. 30, pp 1–7.

Bidwill, J.C., B.U. Edmund, F. Walker, R.P. Marshall and G. Fulford 1851, Resolutions at a meeting of magistrates held at Maryborough, 19 February, CS reference number 51/02577, QLS reel A2.23, pp 847–9, https://bit.ly/2J80OmT.

Breckwoldt, R. 1988, *A very elegant animal the dingo*, Angus & Robertson, Sydney.

Bromiley, R. 1937, The Aboriginal name for Pialba, *Maryborough Chronicle*, 26 August, p 5, https://bit.ly/3nT870G.

Brooks, D. 2016, *Derrida's breakfast*, Brandl and Schlesinger, Blackheath NSW.

Brown, E. 1998, 'Eliza Fraser: an historical record' in (eds) I.J. McNiven, L. Russell and K. Schaffer, *Constructions of colonialism: perspectives on Eliza Fraser's shipwreck*, Leicester University Press, London and New York.

Bryant, C. 2012, Dingoes on Fraser Island are still suffering, SFID media release, 8 September, https://bit.ly/2KGeDcT.

Bryant, C. 2013, Fraser Island dingo incident reports, SFID media release, 4 February, https://bit.ly/2UXVG7d.

Bryson, J. 1986 [1985], *Evil angels*, Penguin, Ringwood VIC.

Burns, L. and P. Howard 2003, When wildlife tourism goes wrong: a case study of stakeholder and management issues regarding dingoes on Fraser Island, Australia, *Tourism management: research, policies, practice*, vol. 24, no. 6, pp 699–712, doi: 10.1016/S0261-5177(03)00146-8.

Cairns, K.M. and A.N. Wilton 2016, 'New insights on the history of canids in Oceania based on mitochondrial and nuclear data', *Genetica*, no. 144, pp 553–65.

Cairns, K., S. Brown, B.N. Sacks and J.W.O. Ballard 2017, 'Conservation implications for dingoes from the maternal and paternal genome: multiple populations, dog introgression, and demography.' *Ecology and evolution*. pp 1–21, doi: 10.1002/ece3.3487.

Cairns, K.M., B.J. Nesbitt, S.W. Laffan, M. Letnic, M.S. Crowther 2020, Geographic hot spots of dingo genetic ancestry in southeastern Australia despite hybridisation with domestic dogs, *Conservation genetics*, no. 21, 77–90, https://bit.ly/378Ep12.

Carter, J., A. Wardell-Johnson and C. Archer-Lean 2017, Butchulla perspectives on dingo displacement and agency at K'gari-Fraser Island, Australia, *Geoforum* no. 85, pp 197–205, doi: https://bit.ly/2UXOwA7.

Chamberlain, L. 1990, *Through my eyes: an autobiography*, William Heinemann, Port Melbourne VIC.

Channel 5 2001, *When dingoes attack*, directed and filmed by Scott Tibbles, produced by Wendy McLean, Available Light Productions, Bristol.

Chewings, C. 1936, *Back in the stone age: the natives of central Australia*, Angus & Robertson, Sydney.

Clarke, K.W. and C.M. Trim 2014, *Veterinary anaesthesia* 11/e, Saunders Elsevier, Edinburgh.

Clendinnen, I. 1999, Inside the contact zone, part 2, Boyer lecture 5, *True stories*, ABC Radio National, 12 December.

Works cited

Code, L. 2008, Review of Miranda Fricker, *Epistemic injustice: power and the ethics of knowing, Notre Dame Philosophical Reviews*, University of Notre Dame, South Bend IN, https://bit.ly/3m9O8dG.

Coppinger, R. and M. Feinstein 2015, *How dogs work*, University of Chicago Press, Chicago and London.

Corbett, L.K. 1988, Social dynamics of a captive dingo pack: population regulation by dominant female infanticide, *Ethology*, no. 78, pp 177–98.

Corbett, L.K. 1995, *The dingo in Australia and Asia*, University of NSW Press, Sydney.

Corbett, L.K. 1998, *Management of dingoes on Fraser Island*, Report for Queensland Department of Environment by ERA Environmental Services, no place.

Corbett, L.K. 2001, 'The conservation status of the Dingo *Canis lupus dingo* in Australia, with particular reference to New South Wales: threats to pure Dingoes and potential solutions' in C.R. Dickman and D. Lunney (eds), *A symposium on the dingo*, Royal Zoological Society of New South Wales, Mosman NSW.

Corbett, L.K. 2004, Dingo in C. Sillero-Zubiri, M. Hoffmann and D.W. Macdonald (eds) *Canids: Foxes, Wolves, Jackals and Dogs: Status Survey and Conservation Action Plan*, International Union for Conservation of Nature and Natural Resources/Species Survival Commission, Canid Specialist Group, Gland, Switzerland and Cambridge, UK https://bit.ly/3fzHJ8Z.

Cripps, S. 2017, 'Government interim conservation order lifted from Hinchinbrook shire', *North Queensland Register*, 11 September, https://bit.ly/3m7jO3j.

Crowther, M.S., M. Fillios, N. Colman and M. Letnic 2014, An updated description of the Australian dingo (*Canis dingo* Meyer 1793), *Journal of zoology*, doi:10.1111/jzo.1213.

Curtis, J. 1838, *Shipwreck of the Stirling Castle*, George Virtue, London.

Daisley, S. 2015, *Coming rain*, Text Publishing, Melbourne.

Davison, F.D. 1983 [1946], *Dusty*, Angus & Robertson, Pymble NSW.

Dawson, N. 2002, Lindy Chamberlain: a personal visual response, Doctor of creative arts thesis, Faculty of Creative Arts, University of Wollongong, http://ro.uow.edu.au/theses/921.

Despret, V. 2016 [2012], *What would animals say if we asked the right questions?* tr. B. Buchanan, University of Minnesota Press, Minneapolis MS and London, e-book.

Director of National Parks 2012, *Uluru–Kata Tjuta National Park knowledge handbook*, Australian Government, no place, https://bit.ly/3fzHZot.

Dixon, J.M. and L. Huxley (eds) 1985, *Donald Thomson's mammals and fishes of northern Australia*, Nelson, Melbourne.

Donaldson, S. and W. Kymlicka 2011, *Zoopolis: a political theory of animal rights*, Oxford University Press, Oxford.

Donaldson, S. and W. Kymlicka 2016, 'Between wildness and domestication: rethinking categories and boundaries in response to animal agency' in B. Bovenkerk and J. Keulartz (eds) *Animal ethics in the age of humans: blurring boundaries in human–animal relationships*, Springer, New York.

Duncan-Kemp, A.M. 1933, *Our sandhill country: nature and man in south-western Queensland*, Angus & Robertson, Sydney.

Dutton, G. 1967, Everard Ranges, *Poems Loud and Soft*, Cheshire, Melbourne.

Dwyer, F. 2015, Interview with Rowena Lennox, Hervey Bay, 27 November.

Ecosure 2012, *Fraser Island dingo management strategy review*, report to Department of Environment and Heritage Protection, Ecosure, West Burleigh QLD. According to a citation in Allen et al. 2015, this report was authored by B.L. Allen, J. Boswell and K. Higginbottom. Their names do not appear on the report.

Ecosure 2013, *Fraser Island dingo conservation and risk management strategy*, Queensland Department of Environment and Heritage Protection, no place, https://bit.ly/39kHzBD.

Editors of the Encyclopaedia Britannica 2020, C. Lloyd Morgan, *Encyclopaedia Britannica*, Encyclopaedia Britannica Inc, https://bit.ly/33gxD88.

Enright, A. 2018, The genesis of blame, *London review of books*, 8 March, pp 6–8.

Essler, J.L., S. Marshall-Pescini and F. Range 2017, Domestication does not explain the presence of inequity aversion in dogs, *Current biology*, no. 27, pp 1861–5, https://bit.ly/364axDv.

ESVCE (European Society of Veterinary Clinical Ethology) 2017, 'ESVCE position statement on electronic training devices', September, https://bit.ly/2KIFFAx.

Evans, R. 2008, Done and dusted, *Griffith review 21 – hidden Queensland*, https://bit.ly/3kYbaCY.

Evans, R. and J. Walker 1977 'These strangers, where are they going?' Aboriginal–European relations in the Fraser Island and Wide Bay region 1770–1905 in P.K. Lauer (ed.) *Fraser Island: occasional papers in anthropology no. 8*, University of Queensland, St Lucia, pp 39–105.

Farlex 2003–2017, *The free dictionary*, Hemorrhage, https://bit.ly/2IYEFYw.

FIDO (Fraser Island Defenders Organisation) 2001, *Moonbi 100*, FIDO, Bald Hills QLD, http://fido.org.au/moonbi/moonbi100.html.

Field, D. 2008, Dingo fence on Fraser Island criticised, *The world today*, ABC radio, 6 May, transcript, https://ab.co/3m5NtKl.

Works cited

Filios, M. and Tacon, P. 2016, Who let the dogs in? A review of the recent genetic evidence for the introduction of the dingo to Australia and the implications for the movement of people, *Journal of archaeological science: reports*. http://dx.doi.org/10.1016/j.jasrep.2016.03.001.

Fleming, P. and Ballard, G. 2013, 'Dingoes, dogs and the feral identity', *The conversation*, 18 February, https://bit.ly/3nVpcqR.

Foucault, M. 1995, *Discipline and punish: the birth of the prison*, 2nd edn, tr. A. Sheridan, Vintage, New York.

Fowell, N. 1788, Letter received by John Fowell, State Library of NSW, SAFEMLMSS 4895/1/IE1610008, LF1610089, 12 July 1788, https://bit.ly/3fwdwaU.

Gardner, F. 2015, Heritage recommendation 602393, Original Maryborough town site (extension), Queensland Department of Environment and Heritage Protection, https://bit.ly/376AL7F.

Giles, E. 1986 [1889], *Australian twice traversed*, vol. 2, Doubleday, Lane Cove NSW.

Gollan, K. 1984, The Australian dingo: in the shadow of man, in M. Archer and G. Clayton (eds) *Vertebrate zoogeography and evolution in Australasia*, Hesperian Press, Carlisle, pp 921–7.

Gould, R.A. 1969, Subsistence behaviour among the western desert Aborigines of Australia, *Oceania*,
vol. 39, no. 4, pp 253–74.

Grant, A. n.d. Place name glossary, Scots words and place names (SWAP), University of Glasgow, https://bit.ly/3nWadgh.

Gunn, I. 2011, Death of the Fraser Island dingo, *The conversation*, 27 April, https://bit.ly/2V01Xzt.

Hadley, J. 2015, *Animal property rights: a theory of habitat rights for wild animals*, Lexington Books, Lanham MD.

Hamilton, A. 1972, Aboriginal man's best friend? *Mankind*, vol. 8, no. 4, pp 287–95.

Hayden, B. 1975, Dingoes: pets or producers? *Mankind*, vol. 10, no. 1, pp 11–15.

Heffernan, J. 2001, *Chips*, Scholastic, Gosford NSW.

Horowitz, A. 2010, *Inside of a dog: what dogs see, smell and know*, Simon & Schuster, London.

ISV (International Student Volunteers) 2018, website of International Student Volunteers, Dominican Republic, https://bit.ly/3pZsPhk.

Johnson, C.N. 2006, *Australia's mammal extinctions: a 50,000 year history*, Cambridge University Press, Cambridge.

Johnson, C.N., J.L. Isaac and D.O. Fisher 2007, 'Rarity of a top predator triggers continent-wide collapse of mammal prey: dingoes and marsupials in

Australia', *Proceedings of the Royal Society B*, no. 274, pp 341–6, doi:10.1098/rspb.2006.3711.

Jones, R. 1970, Tasmanian Aborigines and dogs, *Mankind*, vol. 7, no. 4, pp 256–71.

Jurox 2015, Valabarb euthanasia solution: safety data sheet. https://bit.ly/39dPm3W.

Kafka, F. 1946 [1933], *Investigations of a dog*, tr. W. and E. Muir, Secker and Warburg, London.

Kafka, F. 2017, *Investigations of a dog and other creatures*, tr. M. Hoffman, New Directions, New York.

Knox, E. 2003, Patience in D. Adelaide (ed.) *Acts of dog*, Vintage, Sydney, pp 109–36.

Kolig, E. 1978, 'Aboriginal dogmatics: canines in theory, myth and dogma', *Bijdragen tot de Taal-, Land- en Volkenkunde*, vol. 134, no. 1, pp 84–115.

Kopp, S. 2018, Personal communication with Rowena Lennox, email, 18 May.

Koungoulos, L. 2020, 'Dingo archaeology: a 3D look into the history of our native dog', Archaeology, museums and heritage seminar series, University of Sydney, 28 February.

Lauer, P.K. 1977, 'Report of a preliminary ethnohistorical and archaeological survey of Fraser Island' in P.K. Lauer (ed.) *Fraser Island: occasional papers in anthropology no. 8*, University of Queensland, St Lucia, pp 1–38.

Layton, R. 1989, *Uluru: an Aboriginal history of Ayers Rock*, Aboriginal Studies Press, Canberra.

Lee, L. 2006a, Course introduction and nomenclature, Anesthesiology VMED 7412, Centre for Veterinary Healthy Sciences, Oklahoma State University, 10 January, https://bit.ly/37asVdp.

Lee, L. 2006b, Pharmacology – intravenous anesthetic agents and dissociatives, Anesthesiology VMED 7412, Centre for Veterinary Healthy Sciences, Oklahoma State University, 24 January, https://bit.ly/37at8NJ.

Lennox, R. 2013, Head of a dog, *Southerly*, vol. 73, no. 3, pp 212–26.

Lennox, R. 2014, Apex predators, *Meanjin*, vol. 73, no. 3, pp 6-9.

Lennox, R.J., A.J. Gallagher, E.G. Ritchie and S.J. Cooke 2018, Evaluating the efficacy of predator removal in a conflict-prone world, *Biological conservation*, no. 224, pp 277–89, doi.org/10.1016/j.biocon.2018.05.003.

Leonard, J.A., J.E. Echegaray, E. Randi and C. Vila 2014, Impact of hybridization with domestic dogs on the conservation of wild canids in M.E. Gompper (ed.) *Free-ranging dogs and wildlife conservation*, Oxford University Press, Oxford.

Letnic, M., E.G. Ritchie and C.R. Dickman 2012, 'Top predators as biodiversity regulators: the dingo *Canis lupus dingo* as a case study', *Biological reviews, no. 87*, pp 390–413.

Lynch, D. and G. Lunney 1996, *Maryborough: a rare old town sketched in time*, True Blue Books, Maryborough.

Macdonald, H. 2014, *H is for hawk*, Jonathan Cape, London.

Martin, G.L. 2001a, Dingoes savage boy, nine, to death, *The Telegraph* (UK), 1 May, https://bit.ly/3o1lkVr.

Martin, G.L. 2001b, Grandfather tells of dingo horror, *The Telegraph* (UK), 2 May, https://bit.ly/2Kwk0Lw.

Maryborough portside. 2006–2018, Maryborough port website, https://bit.ly/3l0EWqN.

Mattei, U. 2016, Forum response 'The new nature', *Boston Review*, 11 January.

McNiven, I.J., L. Russell and K. Schaffer (eds) 1998, *Constructions of colonialism: perspectives on Eliza Fraser's shipwreck*, Leicester University Press, London and New York.

Meehan, B., R. Jones and A. Vincent 1999, 'Gulu-kula: dogs in Anbarra society, Arnhem Land', *Aboriginal history*, vol. 23, pp 83–106.

Meggitt, M.J. 1965, 'The association between Australian Aborigines and dingoes' in A. Leeds and A.P. Vayda (eds) *Man, culture and animals: the role of animals in human ecological adjustments*, American Association for the Advancement of Science, Washington DC.

Merriam-Webster 2017, 'Thready pulse', https://bit.ly/3pUeSRP.

Meston, A. 1905, *Report on Fraser Island*, Queensland Legislative Assembly, Brisbane.

Michael, P. 2019, Dingo and billy goat in fatal face-off on survivor island, *The Courier Mail*, 29 June.

Miller, O. 1994, *Legends of Fraser Island*, Rigby Heinemann, Melbourne.

Miller, O. 1998, K'gari, Mrs Fraser and the Butchulla oral tradition in I.J. McNiven, L. Russell and K. Schaffer (eds) *Constructions of colonialism: perspectives on Eliza Fraser's shipwreck*, Leicester University Press, London and New York, pp 28–36.

Mitchell, T.L. 1965 [1839], *Three expeditions into the interior of eastern Australia*, vol. 2, Libraries Board of South Australia, Adelaide.

Moreton Bay Courier 1847, 6 November, p 2. http://trove.nla.gov.au/newspaper/page/540877.

Moreton-Robinson, A. 2004, 'Whiteness, epistemology and Indigenous representation' in A. Moreton-Robinson (ed), *Whitening race: essays in social and cultural criticism*, Aboriginal Studies Press, Canberra.

Morling, T. 1987, Royal commission of inquiry in the Chamberlain convictions, Report, Commonwealth parliamentary papers, Canberra, vol. 15, paper 192, https://bit.ly/360NwkV.

Morris, E. 2012, *Inquest into the death of Azaria Chantel Loren Chamberlain* [2012] NTMC 020, Coroners Court, Darwin.

Mountford, C.P. 1981 [1948], *Brown men and red sand: journeyings in wild Australia*, Angus & Robertson, London and Sydney.

Murray, L. 1997, 'The sand dingoes', *Subhuman redneck poems*, Farrar, Straus and Giroux, New York.

Nabokov, V. 1996 [1976 English, 1925 Russian], 'A guide to Berlin' in D. Nabokov (ed.), *The stories of Vladimir Nabokov*, London, Orion.

NDPRP (National Dingo Protection and Recovery Program) 2016, 'Fraser Island report', *NDPRP magazine*, vol. 4, no. 1, pp 40–7.

Newsome, A.E. and L.K. Corbett 1985, 'The identity of the dingo III: the incidence of dingoes, dogs and hybrids and their coat colours in remote and settled regions of Australia', Australian Journal of Zoology, no. 33, 363–73.

Newsome, T.M. 2014, Makings of icons: Alan Newsome, the red kangaroo and the dingo, *Historical records of Australian science*, vol. 25, no. 2, pp 153–71.

Norton, C. 2012a, Tourist's amazing escape, *Gladstone Observer*, 30 July.

Norton, C. 2012b, Tourist tells of dingo face off, *Gladstone Observer*, 1 August.

Novak, D. 2015, Interview with Rowena Lennox, Eurong QLD, 26 November.

NSW DPI (New South Wales Department of Primary Industries). 2012, *New South Wales wild dog management strategy 2012–2015,* NSW DPI, Orange NSW.

O'Neill, A.J. 2002, *Living with the dingo,* Envirobook, Annandale NSW.

O'Neill, A.J., K.M. Cairns, G. Kaplan and E. Healy 2016, Managing dingoes on Fraser Island: culling, conflict and an alternative, *Pacific conservation biology*, https://bit.ly/3pZuH9Q.

Oskarsson, M.C.R., C.F.C. Klütsch, U. Boonyaprakob, A. Wilton, Y. Tanabe and P. Savolainen 2011, Mitochondrial DNA data indicate an introduction through Mainland Southeast Asia for Australian dingoes and Polynesian domestic dogs, *Proceedings of the Royal Society B*, doi: 10.1098/rspb.2011.1395.

Pang, J.F., C. Kluetsch, X.J. Zou, A.B. Zhang, L.Y. Luo, H. Angleby, A. Ardalan, C. Ekström, A. Sköllermo, J. Lundeberg, S. Matsumura, T. Leitner, Y.P. Zhang and P. Savolainen 2009, 'mtDNA data indicate a single origin for dogs south of Yangtze River, less than 16,300 years ago, from numerous wolves' *Molecular Biology and Evolution,* vol. 26, no. 12, pp 2849–64, doi: 10.1093/molbev/msp195.

Parker, M.A. 2006, Bringing the dingo home: discursive representations of the dingo by Aboriginal, colonial and contemporary Australians, PhD thesis, School of English, journalism and European languages, University of Tasmania.

Parkhurst, J. 2010, *Vanishing icon: the Fraser Island dingo*, Grey Thrush Publishing, St Kilda VIC.

Parkhurst, J. 2012, *Vanishing icon: the Jennifer Parkhurst Fraser Island dingo story*, https://bit.ly/2V1Jr9O.

Parkhurst, J. 2015a, *The Butchulla First Nations people of Fraser Island (K'gari) and their dingoes*, Australian Wildlife Protection Council, Melbourne.

Parkhurst, J. 2015b, Interview with Rowena Lennox, Rainbow Beach and K'gari, 18–19 May.

Parkhurst, J. 2015c, Personal communication with Rowena Lennox, email, 27 August.

Parkhurst, J. 2015d, Interview with Rowena Lennox, telephone, 29 August.

Parkhurst, J. 2018, Personal communication with Rowena Lennox, telephone conversation, 19 June.

Parr, W.C.H., L.A.B. Wilson, S. Wroe, N.J. Colman, M.S. Crowther, M. Letnic 2016, Cranial shape and the modularity of hybridization in dingoes and dogs; hybridization does not spell the end for native morphology, *Journal of evolutionary biology*, no. 43, 171–87, doi 10.1007/s11692-016-9371-x

Philip, J. 2016, Personal communication with Rowena Lennox, email, 21 July.

Philip, J. 2017a, International travels of the Australian Canis dingo part 1, blog post, Smithsonian Institution Archives, 13 June, https://siarchives.si.edu/blog/tag/dingo.

Philip, J. 2017b, Representing the dingo: an examination of dingo–human encounters in Australian cultural and environmental heritage, PhD thesis, School of ecosystem management, University of New England, Armidale NSW.

Pierotti, R. and B.R. Fogg 2017, *The first domestication: how wolves and humans coevolved*, Yale University Press, New Haven.

Pratt, M.L. 1991, 'Arts of the contact zone', *Profession*, Modern Language Association

Probyn-Rapsey, F. 2015, 'Dingoes and dog-whistling: a cultural politics of race and species in Australia', *Animal studies journal*, vol. 4, no. 2, pp 55–77, http://ro.uow.edu.au/asj/vol4/iss2/4.

Probyn-Rapsey, F. 2016, Five propositions on ferals, *Feral feminisms*, no. 6, https://bit.ly/33gozjK.

Prowse, A.A., C.N. Johnson, P. Cassey, C.J.A. Bradshaw and B.W. Brook 2015, Ecological and economic benefits to cattle rangelands of restoring an apex predator, *Journal of applied ecology*, vol. 52, no. 2, pp 455–66, doi: org/10.1111/1365-2664.12378.

Purcell, B. 2010, *Dingo* CSIRO Publishing, Collingwood VIC.

QPWS (Queensland Parks and Wildlife Service) 2015, *The dingoes of Fraser Island (K'gari): safety and information guide*, State of Queensland, Department of National Parks, Sport and Racing, no place.

QPWS (Queensland Parks and Wildlife Service) 2017, *The dingoes (wongari) of K'gari (Fraser Island): safety and information guide*, State of Queensland, Department of National Parks, Sport and Racing, no place, https://bit.ly/ 3nXFEqK.

Queensland government 2014, Dingo management on Fraser Island, Department of National Parks, Sport and Racing website, https://bit.ly/2KElL9I.

Queensland government 2015, Dingo incidents lead to humane destruction, Press release issued by Department of National Parks, Sport and Racing, 17 August. https://bit.ly/33g0GZo.

Queensland government 2017–2018, Department of National Parks, Sport and Racing – disclosure log entries, Department of Environment and Science website, https://bit.ly/3l8WS2d.

Ripple, W.J., J.A. Estes, R.L. Beschta, C.C. Wilmers, E.G. Ritchie, M. Hebblewhite, J. Berger, B. Elmhagen, M. Letnic, M.P. Nelson, O.J. Schmitz, D.W. Smith, A.D. Wallach and A.J. Wirsing 2014, 'Status and ecological effects of the world's largest carnivores' *Science*, no. 343, article no. 1241484.

Ritchie, E.G., C.J.A. Bradshaw, C.R. Dickman, R. Hobbs, C.N. Johnson, E.L. Johnston, W.F. Laurence, D. Lindenmayer, M.A. McCarthy, D.G. Nimmo, H.H. Possingham, R.L. Pressey, D.M. Watson and J. Woinarski 2013, Continental-scale governance failure will hasten loss of Australia's biodiversity, *Conservation Biology*, vol. 27, no. 6, pp 1133-5, doi: 10.111/ cobi.12189.

Robson, F. 2013, The dingo woman, *Sydney Morning Herald*, 13 April.

Rose, D.B. 2000 [1992], *Dingo makes us human: life and land in Australian Aboriginal culture*, Cambridge University Press, Cambridge.

Rose, D.B. 2011, *Wild dog dreaming: love and extinction,* University of Virginia Press, Charlottesville NC and London.

Rose, D.B. 2013, Slowly – writing into the Anthropocene, *TEXT* special issue 20, October.

Roughsey, D. 1973, *The giant devil dingo*, Collins, Sydney.

Ryan, J.S. 1964, Plotting an isogloss – the location and types of Aboriginal names for native dog in New South Wales, *Oceania*, vol. 35, pp 111-23.

Saint-Exupéry, A. 2005 [1943] *The little prince*, tr. R. Howard, Egmont, London.

Savolainen, P., T. Leitner, A.T. Wilton, E. Matisoo-Smith and J. Lundeber 2004, A detailed picture of the origin of the Australian dingo, obtained from the study of mitochondrial DNA, *Proceedings of the National Academy of Sciences of the United States of America*, vol. 101, no. 33, pp 12387–90, doi: 10.1073/ pnas.0401814101.

Works cited

SFID (Save Fraser Island Dingoes) 2015, *The dingo – friend or foe?* Forum held at Hervey Bay Community Centre, 17 May. https://www.youtube.com/watch?v=jwcghHGAF7c.

Schaffer, K. 1995, *In the wake of first contact: the Eliza Fraser stories*, Cambridge University Press, Cambridge.

Schlesinger, L. 2018, SeaLink buys Fraser Island Kingfisher Bay resorts for $43m, *Australian Financial Review*, 21 February, https://bit.ly/3l0dol7.

Schwartz, D. 2016a, Death row dingoes set to be the environmental saviour of Great Barrier Reef's Pelorus Island, *ABC Landline*, 23 July, https://ab.co/2JfkCEY.

Schwartz, D. 2016b, 'Death-row dingoes' plan to eradicate goats axed by Queensland government to save vulnerable curlew, *ABC news*, 18 August, https://ab.co/2V2NWB7.

Shakespeare, W. n.d., *Hamlet, prince of Denmark* in *The complete works of William Shakespeare*, The Literary Press Limited, London.

Simper, E. 2010, Discovery of jacket vindicated Lindy, *The Australian*, 14 August.

Skelly, K. 2015, Interview with Rowena Lennox, Hervey Bay, 25 November.

Skinner, L.E. 1974, The search for the *Sea Belle* castaways on Fraser Island, *Queensland heritage*, vol. 2,
no. 10, pp 3–14.

Smith, B. 2015, Biology and behaviour of the dingo, in B. Smith (ed.) *The dingo debate: origins, behaviour and conservation*, CSIRO Publishing, Clayton South VIC.

Smith, B. and Appleby, R. 2018, Promoting human–dingo co-existence in Australia: moving toward more innovative methods of protecting livestock rather than killing dingoes (*Canis dingo*), *Wildlife research*, no. 45, pp 1–15, https://doi.org/10.1071/WR16161.

Smith, S. 1978, *Stevie Smith: selected poems,* J. MacGibbon (ed.), Penguin, Harmondsworth.

Smyth, R.B. 1972 [1876], *The Aborigines of Victoria*, vol. 1, John Currey, O'Neil, Melbourne.

Stephens, D., A.N. Wilton, P.J.S. Fleming, O. Berry 2015, 'Death by sex in an Australian icon: a continent-wide survey reveals extensive hybridizaton between dingoes and domestic dogs', Molecular Ecology, no. 24, 5643–56.

Tench, W. 1789, *A narrative of the expedition to Botany Bay: with an account of New South Wales, its productions, inhabitants &c.*, no publisher, London.

The Triffids, 1986, 'Wide open road' (song), *Born sandy devotional* (album), White Hot/Mushroom, Domino records, recorded London 1985, https://www.youtube.com/watch?v=7N5akOOlGTI.

The Who. 1978, 'Who are you?' (song), *Who are you?* (album), Polydor/MCA, London.

Tiffany, C. 2012, *Mateship with birds*, Pan Macmillan, Sydney.

Tindale, N.B. 1974, *Aboriginal tribes of Australia: their terrain, environmental controls, distribution limits, and proper names*, University of California Press, Berkeley, Los Angeles, London.

Tucker, M. (ed.) 1995, *The Duke Ellington reader*, Oxford University Press, New York.

University of Queensland and FAIMS (Field acquired information management systems project) 2014, Original Maryborough town site, Australian e-heritage portal, https://bit.ly/2KGjk6v.

Victoria DEPI (Department of Environment and Primary Industries) 2013, *Action plan for managing wild dogs in Victoria 2014–2019*, State Government of Victoria, DEPI, Melbourne.

Virbac 2015, website of Virbac UK, Bury St Edmunds, https://uk.virbac.com/zoletil.

Vogler, S. 2016, Self-culling dingoes could be conservation model, *The Australian*, 26 July, https://bit.ly/3l8RjB0.

Walker, C. 2013, What led to Fraser Island dingo Inky's death, *Fraser Coast Chronicle*, 28 January.

Walker, F. 1851, Letter to colonial secretary, 6 July, CS reference number 51/07537, received 2 August 1851, QLS reel A2.23, pp 831–2, https://bit.ly/2J80OmT.

Walker, F. 1852. Letter reporting result of attempt by Native Police to apprehend blacks accused of felony, & who took refuge on Fraser's Island, 5 January, CS reference number 52/00715, received 22 January 1852, QLS reel A2.23, pp 820–30. https://bit.ly/2J80OmT.

Wallach, A.D. 2011, *Reviving ecological functioning through dingo restoration*, PhD thesis, School of earth and environmental sciences, University of Adelaide.

Wallach, A.D, Bekoff, M., Nelson, M.P. and Ramp, D. 2015, 'Promoting predators and compassionate conservation', *Conservation biology*, vol. 0, no. 0, pp 1–4, doi:10.1111/cobi.12525.

Warner, M. 2014, Story-bearers, *London review of books*, vol. 36, no. 8, pp 19–20.

Watson, L. 2015, Interview with Rowena Lennox, Toolern Vale VIC, 16 July.

Weisse, A. and Ross, A. 2017, 'Managing a contested cultural heritage place on K'gari (Fraser Island), Queensland, Australia', *Archaeology in Oceania*, vol. 52, 149–60, doi 10.1002/arco.5130.

White, P. 1995 [1966], *The solid mandala*, Vintage, London.

White, T.H. 1963 [1951], *The goshawk*, Penguin, Harmondsworth.

Wicks, S., K. Mazur, P. Please, S. Ecker and B. Buetre 2014, *An integrated assessment of the impact of wild dogs in Australia*, Research report no. 14.4,

Works cited

Australian Bureau of Agricultural and Resource Economics and Sciences (ABARES), Canberra.

Williams, F. 1982, *Written in sand: a history of Fraser Island,* Jacaranda Press, Milton QLD.

Williams, F. 2002, *Princess K'gari's Fraser Island,* Fred Williams, Brisbane.

Wilson, N. 2001 *A dogger's life,* Aussie Outback Publishing, Bendigo.

WoolProducers Australia. 2014, *National wild dog action plan: promoting and supporting community-driven action for landscape-scale wild dog management,* final draft, WoolProducers Australia, Barton ACT.

Wright, A. 2016, What happens when you tell somebody else's story? *Meanjin,* vol. 75, no. 4. Available at https://bit.ly/3kYhwCk.

Wright, J. 1946, Trapped dingo, *The moving image, Meanjin,* Melbourne.

Wright, J. 2015 Playing with fire, *Griffith review 47 – looking west,* https://griffithreview.com/articles/playing-fire/.

Youlden, H. 1853, Shipwreck in Australia, *Knickerbocker,* vol. 41, no. 4, April.

Young, N.H. 1989, *Innocence regained: the fight to free Lindy Chamberlain,* Federation Press, Annandale NSW.

Young, S. 2012, *Craving earth: understanding pica – the urge to eat clay, starch, ice and chalk,* Columbia University Press, New York.

Acknowledgments

I am indebted to many people who helped to make this book possible. Jennifer Parkhurst, Kay Skelly, Dan Novak, Linda Behrendorff and Finn Dwyer generously agreed to let me to interview them and gave me permission to use our interviews; Jennifer Parkhurst and Dan Novak kindly allowed me to use their photographs; Debra Adelaide had faith in this project, and her support and insights while she supervised my doctorate of creative arts at the University of Technology Sydney made writing this book such a pleasure; John Dale, my alternative supervisor at UTS, always boosted my confidence; Grace Moore and Martin Thomas were perceptive examiners of my thesis. Karin and Malcolm Kilpatrick at Save Fraser Island Dingoes, and Ross Belcher and Linda Behrendorff at Queensland Parks and Wildlife Service assisted my research in crucial ways. James Fawcett, Adam O'Neill, Mark Robinson, Arian Wallach and Lyn Watson shared their insights with me; Clare Archer-Lean reviewed the manuscript – and had my back; Justine Philip and Amanda Stuart were fellow dingo travellers; Simone Ford proofread my thesis; Laurie Whiddon drew the beautiful map. I acknowledge the work of Deborah Bird-Rose, who was an interlocutor for the Aboriginal people of Yarralin in the Northern Territory of Australia and who wrote about dingoes with such grace and wisdom. So many have contributed to this book. I hope those whose names I have omitted here will please forgive me.

The Faculty of Arts and Social Sciences and the Graduate Research School at UTS provided a home and support for my research and writing; the Australian Government's Research Training Program supported the research; members and staff of the Australasian Animal Studies Association, the Butchulla Prescribed Body Corporate, Fisher Library at the University of Sydney, Fraser Explorer Tours, the National Museum of Australia, Special Collections in the Barr Smith Library at the University of Adelaide, and the Blake Library at the University of Technology Sydney, supported and contributed to the research in various capacities. Organisers of, participants in and universities that hosted various animal studies, compassionate conservation, creative writing, environmental history, geography, literature, multispecies ethnography and speculative ethology symposia, workshops, reading groups and conferences enriched my research and made it much more fun.

Big thanks to Melissa Boyde and Fiona Probyn-Rapsey of the Animal Publics series for taking on *Dingo Bold* and to the wonderful people at Sydney University Press for the care with which they have taken me through the publishing process and turned the manuscript into a book: Agata Mrva-Montoya, Denise O'Dea and Jo Butler – I couldn't have asked for more sympathetic, attentive editing.

I thank the editors of the following publications for including earlier versions of parts of this book, and essays and interviews that grew from the research, and I am grateful for the support and recognition of a Queensland Writing Fellowship and an Australasian Association of Writing Programs creative writing prize: Gillian Dooley and Danielle Clode (eds), *The first wave: exploring early coastal contact history in Australia*, Wakefield Press, Adelaide, 2019 ('Incessant: dingoes and waves of contact on K'gari'); *Authorised theft: writing, scholarship, collaboration*, refereed proceedings of the 21st conference of the Australasian Association of Writing Programs, Canberra, 2016 ('Coolooloi', winner of the Creative Stream of the 2016 AAWP Postgraduate Writing Prize); *The big issue – love edition*, no. 570, 7–20 September 2018 ('The space between us'); *Griffith review – perils of populism*, no. 57, 2017 ('Killing Bold: managing the dingoes of Fraser Island', awarded a 2017 *Griffith review* Queensland Writing Fellowship); *Meanjin*, vol. 73, no. 3, 2014 ('Apex predators'); *Southerly – the naked writer*, vol. 73, no. 3, 2013 ('Head of a dog'); Australasian Animal

Studies Association blog, 30 June 2016 ('Safe place (dingoes)'); *Meanjin* blog, 2 December 2013 ('What I'm reading'); *Writers in conversation*, vol. 1, no. 1, 2014 ('Interview with Bill Gammage').

This book is a multivocal work, and I thank the writers who have inspired me and whom I have quoted in the text. I gratefully acknowledge permission to reproduce the following epigraphs: chapter 2 © Stephen Daisley, taken from *Coming Rain*, published by the Text Publishing Company Australia; used with permission of the author; chapter 9 © John Heffernan, taken from *Chips*; used with permission of the author; chapter 11 © Les Murray, taken from 'The sand dingoes' in *Subhuman redneck poems*, has been reproduced by permission of the estate of Les Murray c/o Margaret Connolly & Associates Pty Ltd; chapter 12 from *H is for Hawk* by Helen Macdonald published by Vintage. Reproduced by permission of the Random House Group Ltd Ltd © 2015.

I thank Queensland Parks and Wildlife Service for permission to reproduce the cover photograph by Neil Cambourn © Queensland Government, reproduced with the permission of the Department of Environment and Science, Queensland.

Last but not least, thanks and love to Bryn Schwarzl and Chloe Schwarzl who put up with my fascination with dingoes and gave me wise answers to my questions; to Lyn Lennox who shared her stories, read my work and showed her love for our family in so many ways; to Paul Schwarzl who supported me and my writing so wisely and kindly; to Zefa who inspired me and showed by example how to sleep peacefully; and to Bold for being himself.

Index

Index

www.ingramcontent.com/pod-product-compliance
Lightning Source LLC
Chambersburg PA
CBHW020527270326
41927CB00006B/470